'What goes on in the mind of trophy hunters? What drives them to spend huge sums of money in the pursuit of increasingly rare animals so they can kill them and take parts of their bodies home? Are trophy hunters pre-disposed to kill? Why do they feel the need to possess? Why are they lauded by some and reviled by many? Are they a mutated throwback to the psychology of our hunting ancestors or are they simply out of step with today's evolving, ethical perceptions of wildlife and its values? Professor Geoff Beattie brings his intellect, considerable experience and detailed analysis to bear on an issue that divides humanity. *Trophy Hunting: A Psychological Perspective* is required reading.'

Will Travers OBE, President and Co-Founder, Born Free Foundation

Trophy Hunting

This book explores the psychology of trophy hunting from a critical perspective and considers the reasons why some people engage in the controversial activity of killing often endangered animals for sport.

Recent highly charged debate, reaching a peak with the killing of Cecil the lion in 2015, has brought trophy hunting under unprecedented public scrutiny, and yet the psychology of trophy hunting crucially remains under-explored. Considering all related issues from the evolutionary perspective and 'inclusive fitness', to personality and individual factors like narcissism, empathy, and the Duchenne smiles of hunters posing with their prey, Professor Beattie makes connections between a variety of indicators of prestige and dominance, showing how trophy hunting is inherently linked to a desire for status. He argues that we need to identify, analyse and deconstruct the factors that hold the behaviour of trophy hunting in place if we are to understand why it continues, and indeed why it flourishes, in an age of collapsing ecosystems and dwindling species populations.

The first book of its kind to examine current research critically to determine whether there really is an evolutionary argument for trophy hunting, and what range of motivations and personality traits may be linked to this activity. This is essential reading for students and academics in psychology, geography, business, environmental studies, animal welfare as well as policy makers and charities in these and related areas. It is of major relevance for anyone who cares about the future of our planet and the species that inhabit it.

Geoffrey Beattie is a professor of psychology at Edge Hill University and in recent years a master's supervisor on the Sustainability Leadership Programme at the University of Cambridge and a visiting professor at the Bren School of Environmental Science & Management at the University of California, Santa Barbara.

He earned a First Class Honours degree in psychology from the University of Birmingham and a PhD in psychology from the University of Cambridge. He was awarded the Spearman Medal by the British Psychological Society for 'published psychological research of outstanding merit' and the internationally acclaimed Mouton d'Or for his work in semiotics.

Professor Beattie is both a Chartered Psychologist and a Chartered Scientist. He is also a Fellow of the British Psychological Society, a Fellow of the Royal Society of Medicine, a Fellow of the Royal Society of Arts, and a former President of the Psychology Section of the British Association for the Advancement of Science.

Trophy Hunting

A Psychological Perspective

Geoffrey Beattie

LONDON AND NEW YORK

First published 2020
by Routledge
2 Park Square, Milton Park, Abingdon, Oxon OX14 4RN

and by Routledge
52 Vanderbilt Avenue, New York, NY 10017

Routledge is an imprint of the Taylor & Francis Group, an informa business

British Library Cataloguing-in-Publication Data
A catalogue record for this book is available from the British Library

Library of Congress Cataloging-in-Publication Data
A catalog record has been requested for this book

ISBN: 978-0-367-27817-5 (hbk)
ISBN: 978-0-367-27816-8 (pbk)
ISBN: 978-0-429-29798-4 (ebk)

Typeset in Sabon
by Nova Techset Private Limited, Bengaluru & Chennai, India

For my beautiful daughter Zoe, fearless as always in that battle with that strangest, most random and most sinister of enemies ('WAWAW', as we say around here).

Contents

Acknowledgements

I would like to thank the Born Free Foundation, and especially Will Travers, Mark Jones and Nikki Mason, for their support and their encouragement to consider the issue of trophy hunting in a critical fashion to attempt to understand the psychology (or psychologies) behind it.

I would also like to express my gratitude to Edge Hill University, particularly the Vice Chancellor, John Cater, the Pro Vice-Chancellor for Research, George Talbot, and my Head of Department, Rod Nicolson, for providing an excellent (and beautiful) environment for allowing creativity and scholarship to develop and thrive. Finally, I would like to express my sincerest gratitude to Laura McGuire for her help and assistance throughout the course of the research and the writing.

Chapter 1

Introduction

Ethics, emotions, and behaviours

President Theodore Roosevelt and his prey.

In a civilized and cultivated country wild animals only continue to exist at all when preserved by sportsmen. The excellent people who protest against all hunting, and consider sportsmen as enemies of wild life, are ignorant of the fact that in reality the genuine sportsman is by all odds the most important factor in keeping the larger and more valuable wild creatures from total extermination.

Theodore Roosevelt
26th President of the United States
1901–1909

President Theodore (Teddy) Roosevelt was an important advocate of trophy hunting. In his book *African Game Trails: An Account of the African Wanderings of an American Hunter-Naturalist*, Roosevelt described his hunting adventures in East Africa, reporting a kill tally of 512 for himself and his son Kermit. The tally is described in great detail, it included 17 lions (9 shot by Teddy; 8 shot by Kermit), 11 elephants (8 by Teddy; 3 by Kermit) 8 hippopotami (7 by Teddy; 1 by Kermit) and 9 giraffes (7 by Teddy; 2 by Kermit). It seems that Kermit, however, had more luck with the faster animals—the leopards and the cheetahs. Kermit bagged 3 leopards and 7 cheetahs; Teddy didn't bag any.

This safari was described as a 'conservation mission' although Roosevelt expressed considerable pride in his and his son's hunting ability. They kept about a dozen trophies for themselves and donated the rest to the Smithsonian Institution, the world's largest museum, with the aim 'to promote the natural world and science'. A former President of the United States participating in and endorsing trophy hunting lent and continues to lend considerable respectability to the activity, especially with the emphasis on 'conservation'. Roosevelt did, however, stress throughout the personal thrill of hunting big game in the strongest and most visceral of terms: 'it made our veins thrill', he wrote. The fundamental conflict between conservation and the joy of killing is reflected throughout his discourse.

There has been a great deal of highly charged emotional debate and indeed moral outrage about trophy hunting, especially over the past few years (see, for example, Nelson, Bruskotter, Vucetich and Chapron 2016). The act of paying large sums of money to travel to Africa or similar locations to kill certain trophy species, particularly the Big Five—lion, elephant, rhino, buffalo and leopard—for sport, and then to display the carcasses in images which include the (almost invariably smiling) hunter and the means of killing (rifle or crossbow) evokes strong and very powerful emotions.

It evokes strong emotions in those who oppose trophy hunting, who see it as cold-blooded, premeditated murder of majestic animals (analogous, in many people's minds to *serial* killing given that the act is often repeated, finances allowing) with devastating implications for the conservation of certain species of animal, particularly rare species—rarity, of course, being one of the most significant factors that influences the value of the trophy animal alongside the 'charisma' of the species (Johnson et al. 2010).

Trophy hunting also elicits strong emotions in those who support it, who view it as an accomplishment of the highest order, the ultimate test of human skill and endurance in a battle with nature, a natural act, part of the evolutionary cycle, with *positive* implications (they argue—many say contentiously) for conservation, in terms of the fees intended to trickle down into the local economies to support conservation work.

However, a core issue for both positions is what might be called 'the psychology of trophy hunting'. Why do men and women want to pursue this activity in the first place? What drives them? Can we understand the desire or the need to engage in trophy hunting in psychological terms? Is this a primary or a secondary consideration? Does the underlying psychology have implications for all the economic and ethical arguments that surround trophy hunting? Are these economic and ethical arguments nothing more than justifications and rationalisations for forms of behaviour that are primitive and unknown? Or does the hunting of wild animals follow a strict moral and ethical code that has very positive effects for the preservation of wildlife, as Teddy Roosevelt suggested, if we can rid ourselves of the overpowering emotions (disgust, sadness, anger, shame, guilt, perhaps even fear) that may cloud our vision?

These psychological considerations, I will suggest, are central to the debate on trophy hunting, including the ethical discourse around it. They are always present in one form or another, sometimes explicitly but often insidiously and implicitly.

It is worth noting at the outset that there are strong ethical arguments firmly presented by *both* sides of the trophy hunting divide. Trophy hunters argue primarily about the 'naturalness' of hunting for mankind, fixed in time and place, it would seem, by certain evolutionary constraints. In the words of Nils Peterson (2004) 'According to this concept, humans are predators and hunting is the only way for them to enter nature as a participant rather than a spectator....Thus, hunting is "right" because it is a natural human role' (Peterson 2004: 311). It is more than natural, according to some who employ this ethical framework; it also represents 'an honest relationship with nature while most others are deceptive.'

Of course, the argument about naturalness only takes us so far. Many other features of hunter–gatherer societies are also 'natural' in that they are common and regular. Such features represent the established order: the way things are. Consider, for example, the regular and recurring practices of male patriarchy and the subjugation of women that over time appear 'natural'. Critics would say that this makes the naturalness argument somewhat less appealing.

Michael Nelson from Oregon State University and his colleagues (2016) discussed some of the other ethical issues involved in trophy hunting and provided a critical appraisal of each of the standard arguments used. They began by rightly recognising the importance of Aldo Leopold (1933) for establishing how we think and reason in ethical terms about wildlife management. Leopold wrote about 'the art of making land produce sustained annual crops of wild game for recreational use.' Nelson and his colleagues argued that this is still the dominant perspective for viewing wildlife and justifying many forms of hunting, including trophy hunting. Wild animals should be considered as crops to be cultivated and harvested for human recreational use. 'Harvested' is a very frequent and recurrent trope used throughout the discourse of trophy hunters, as we shall see. Nelson and colleagues also pointed out that the arguments and debates about trophy hunting (often about its appropriateness for animal conservation) often 'skirt [rather than strictly address] the broader ethical question sitting at the heart of the controversy. This question, put simply, is: what constitutes a good reason to kill an animal?' (Nelson et al. 2016: 303). Trophy hunters, the researchers argue, most commonly use the ethical theory of consequentialism to justify their actions. This theory holds that the consequence of one's action or practice is the sole basis for judging whether the action is right or wrong (reflected in that well-worn aphorism that the ends justify the means).

However, they point out the shortcomings of this argument. Firstly, they remind us that in many situations, the ends clearly do not justify the means. They use the example of human sex trafficking. They write 'The revenue that could be generated [for philanthropic purposes from such an activity] is not sufficient to override the wrong that is done when we condone human trafficking' (Nelson et al. 2016: 303). The second major shortcoming of consequentialism, they argue, is its tendency to downplay or ignore the importance of *motivation* when attempting to assess the rightness or wrongness of the action in question (as if motivation is irrelevant to ends-and-means arguments).

Motivation is a critical component in many ethical judgments. For example, murder and manslaughter may both be very wrong, but how wrong depends upon judgments about motivation and, of course, judgments about the intentionality of the act (which is intimately bound up with motivation). The motivation to kill an animal like Cecil just to acquire a trophy (with a series of concomitant judgments about the psychological characteristics of someone who would need that kind of trophy to satisfy his or her psychological needs in this way), rather than engaging in the killing to provide a source of food or as a defensive or protective act, would lead to a much more severe (negative) ethical judgment than the same act perpetrated with a different set of underlying motivations. Critics of trophy hunting would (and do) argue that hunting is immoral, not just because it requires intentionally inflicting harm on innocent creatures with morally questionable motivations, but also because of what it tells us about the individuals involved.

In a piece in *The Conversation*, Duclos (2017) calls this 'the objection from character' and further states that 'This argument holds that an act is contemptible not only because of the harm it produces, but because of what it reveals about the actor. Many observers find the derivation of pleasure from hunting to be morally repugnant.' Our inferences about the characteristics of those involved may well influence our ethical decisions in this domain. Thus, considerations about underlying psychological motivation and psychological perceptions of character would seem to be important aspects of the ethical decision-making process when it comes to trophy hunting. Such psychological considerations are not separate from some sort of abstract, context-free ethical framework, which avoids any analysis of the human element in the controversy.

There is, however, another very important psychological aspect to ethical judgments about trophy hunting, in that it can be highly *emotional*. Indeed, evidence of strong emotions in the discourse of those who oppose trophy hunting often attracts considerable criticism from those who support it. They argue that opponents of trophy hunting have their judgment *clouded* (or narrowed or over-ridden) by emotion. They go on to say that, on the one hand, we have 'rational' arguments about wildlife management, conservation and the 'harvesting of certain animals, and on the other a somewhat irrational outpouring of emotion, which just serves to confuse the matter.

Nelson for one does not accept this: he argues based on research in decision-making that emotion may not just be understandable in this and related contexts, but adaptive as well—a *guide* to rational decision-making rather than a hindrance or a substitute (Bechara, Tranel and Damasio 2000; Bechara 2004; Damasio 1994). Indeed, a large body of research in neuroscience on the relationship between emotion and thinking demonstrates that one system, the emotion-based, unconscious automatic system (called System 1 by the Nobel Laureate Daniel Kahneman, 2011) often precedes and directs the other slower, more reflective and conscious system (called System 2 by Kahneman).

The research of Antonio Damasio and his colleagues (Damasio 1994; Bechara et al. 2000) has shown that the emotion-based system focuses attention, has a major effect on what we remember and is more closely linked to behaviour in many situations than are conscious cognitions (see also Beattie and McGuire 2015; Walsh and Gentile 2007). Furthermore, it appears that the emotional system *precedes* activation of any conceptual or reasoning system.

Damasio demonstrated this with a very simple gambling experiment. Sitting in front of the experimental participants are four decks of cards, in their hands they have $2000 to gamble with. Their task is to turn over one card at a time to win the maximum amount of money; with each card the player either wins or loses some money. In the case of two of the decks, the rewards are great ($100) but so too are the penalties. A participant playing either of these two decks for any period of time will end up losing money. On the other hand, if he or she concentrates on selecting cards from the other two decks, the rewards are smaller ($50) but the penalties are also smaller, so the player ends up winning money.

What Damasio found with people playing this game was that after encountering a few losses normal participants generated skin conductance responses (signs of autonomic arousal) before selecting a card from the 'bad deck' and they also started to avoid the decks associated with bad losses. In other words, they showed distinct emotional responses to the bad decks even before they had a conceptual understanding of the nature of the decks and long before they could explain what was going on in this experiment. It seems that they started to avoid the bad decks based on their emotional responses.

Damasio also found that patients with damage to an area of the brain called the ventromedial prefrontal cortex failed to generate skin conductance responses before selecting cards from the bad deck and also did not avoid the decks with large losses. Patients with damage to this part of the brain could not generate the anticipatory skin conductance responses and could not avoid the bad decks even though they conceptually understood the differences in the decks before them. 'The patients failed to act according to their correct conceptual knowledge' according to Bechara et al. (1997: 1294). In other words, Damasio and his colleagues demonstrated that 'in normal individuals, non-conscious biases guide behaviour before conscious knowledge does. Without the help of such biases, overt knowledge may be insufficient to ensure advantageous behavior.' In normal people, activation of the emotional system precedes activation of the conceptual system and we now know the neural connection between these two systems is in the ventromedial prefrontal cortex.

Subsequently, Damasio demonstrated the powerful role of emotions in the generation of moral judgments in that patients with bilateral damage to the ventromedial prefrontal cortex were more likely to choose 'heroic' and highly emotional personally aversive responses when presented with a series of moral dilemmas (Koenigs et al. 2007). Haidt (2001: 181) developed a new model of moral judgment (and evaluative judgment in general) in which *moral judgment* (or *evaluative judgment*) appear automatically and effortlessly in conscious behaviour, but 'moral *reasoning* is an effortful process, engaged in after a moral judgment is made, in which a person searches for arguments that will support an already-made judgment.' In other words, we make our minds up quickly and the arguments presented to us may play little role in our judgment except in the subsequent justification of our behaviours to ourselves or others.

This research explains why *some* campaigns to change behaviour in both public policy and the commercial domain work so well. They target the non-conscious biases head-on. Storey (2008) writes that 'Numerous studies have identified that emotional stimuli make far more effective prompts than purely rational arguments when it comes to changing opinions and provoking a response' (2008: 23). The way that the brain is hard-wired suggests that this might well be the most appropriate strategy. These non-conscious biases affect behaviours long before we understand the significance of the issue that we are acting towards.

If we encounter strong emotions with regard to trophy hunting, we should not necessarily assume that this is 'irrational lunacy' (in Nelson's words) because 'emotional reactions to injustice are normal and healthy, and emotions can be critical for making "good" judgments and decisions' (Nelson et al. 2016: 305). The authors also suggest that the repeated use of this fundamental idea that emotion is the antithesis of rationality to dismiss the emotionally tainted arguments of the opponents of trophy hunting should be considered carefully (especially in the light of the research by Damasio and others).

Rust and Verismo (2015) write in *The Conversation* that 'While it is sad that we sometimes have to resort to killing animals, for conservation, let's not allow emotions to overtake our arguments.' We should perhaps pause to consider the fact that although, on occasion, emotion *can* interfere with decision-making (as everyday experience sometimes teaches us), at other times and in many other domains, emotion may guide and direct cognition in adaptive ways. This theme will reappear throughout this book.

Two tribes

Clearly, there are two very different positions on trophy hunting, both of which have at their cores all kinds of assumptions, some explicit and some implicit, about the psychological factors associated with engaging or not engaging in trophy—cold-blooded psychopathological killers on one side (the view of some opponents of trophy hunting to describe its proponents) versus weak-willed liberals who lack the strength, courage or ability to test themselves in this way on the other (the view of some trophy hunters about their critics). Both groups may suffer from what has been described as the 'false consensus error' by Ross (1977) and others. People suffering from this type of error overestimate the relative commonness of their own behaviours (and support for the behaviours), normalise them, and assume that behaviours different from their own (trophy hunting from the perspective of non-hunters; not engaging in hunting or even able to contemplate it, in the case of the hunters) is more revealing of the underlying *dispositional* characteristics of the individuals involved. This may shed some light on the conflict over trophy hunting.

In a well-known article from the 1970s, Lee Ross (1977) considered how people reasoned about their social worlds. He asked whether people are good intuitive psychologists, as many of us like to think, carefully considering the behavioural data of everyday life, and the antecedents of action to arrive at conclusions about why they behave the way they do. His answer was that we might think of ourselves as rational agents and good observers of people, keen on analysis and reflection, but, we are subject to a whole series of biases that impact on what we see, how we interpret it and the conclusions we draw. He concluded, for example, that we tend to view our own behaviour as being appropriate in any social situation and see behaviour *different* from our own as both inappropriate and more indicative of underlying disposition. We assume that this different behaviour tells us more about the underlying characteristics of the other person. We also often suffer from a false consensus effect believing that our own behaviour is more common than it really is, and that most people share our world views, attitudes, preferences and even our emotional states.

For example, Ross (1977) reported that individuals who suffer from depression reckon that 55.1% of people generally suffer from depression. 'Put more money into mental health' they say, 'it's an epidemic.' Those who do not suffer from depression reckon that only 39.2% suffer from depression. 'It's bad', they say, 'but it's not that bad'.

This has implications for many major societal phenomena that affect us all, such as trophy hunting and climate change. Both domains are clearly linked at the conceptual level in that they have aspects of sustainability at their core—of various animal species in one case and our planet in the other. Ross never considered climate change or trophy hunting, but he studied other destructive and cataclysmic events like beliefs in the possibility of a nuclear war. Those who believed that nuclear war was inevitable reckoned that 58.8% of people generally thought that it would occur. Those who didn't think that nuclear war would occur reckoned that only 31.2% of people thought that it would happen.

The false consensus effect amplifies the fear of nuclear war or complacency about it, depending on one's point of view. If you believe that a nuclear war is inevitable and you also think that most people believe that, your belief may mobilise you towards certain actions, not all of which are benign. When it comes to climate change, there are both believers ('Everybody knows that we're heading towards a catastrophe') and deniers ('What's all the fuss about? There is a small cabal of politically-motivated campaigners out for their own ends'.) Each group then makes more extreme and confident judgments about the dispositions of the other group. Believers in the US are perceived as 'Democrats, liberal elite, left-wing environmentalists, possibly socialist, certainly anti-capitalism, agitators'. The deniers are perceived as 'Republicans, right-wing, conservative, anti-scientific, proponents of religious values, doctrinaire'. It is easy to see how this difference may play out with climate change and with many other issues. The difference is not simply a schism in beliefs; it is a schism bolstered by a false consensus and more extreme thinking by one group about the 'other'.

Leviston and colleagues (2012) asked over 5000 Australians to describe their opinions on climate change based on several choices (e.g., 'don't know', 'not happening', 'a natural occurrence', 'human-induced'). The researchers also asked participants to estimate what proportion of the population fell into each category. They found that over 90% of their participants believed that climate change was happening; just over half (50.4%) accepted that human beings were responsible; only 5% reported that climate change was not happening. Those who didn't think that the climate was changing, however, estimated that over 40% of Australians shared their view—clearly a highly inflated consensus effect.

So where does this false consensus effect come from? The false consensus effect, as Beattie and McGuire (2018) and others have noted, is a form of social bias linked to various cognitive and motivational biases. Firstly, it arises from the necessarily biased samples of our own personal social experiences. Ross argues that we tend to mix with people who share similar views to ourselves, and thus we do not sample representative behaviours, beliefs or attitudes to arrive at an accurate conclusion. When we are asked to think of how common a behaviour or belief is, we draw upon social data that comes rapidly to mind, for example, what friends have just said about climate change or images that they shared on social media (such as iconic images of snowfall in spring versus a stranded polar bear). This also influences our judgments about risks and perceptions that climate change is either very risky or a total hoax. Tversky and Kahneman (1973) called this the 'availability heuristic' in that we base our judgments about probability and risk on data that is easiest to recall. Vivid images on social media will exert powerful effects on our judgment and they outweigh probabilistic statistical information in science reports and in the media.

Secondly, there is a good deal of ambiguity in everyday social life which allows for different interpretations depending on our underlying knowledge, attitudes and beliefs. There is a degree of 'cherry picking' of the evidence that supports our underlying beliefs through a confirmation bias. George Marshall, in his book titled *Don't Even Think About It*, says that this type of interpretation operates at all kinds of levels, even in terms of the perception of the weather. He writes, 'When asked about recent weather in their own area, people who are already disposed to believe in climate change will tend to say it's been warmer. People who are unconvinced about climate change will say it's been colder. Farmers in Illinois, invited to report their recent weather experiences, emphasised or played down extreme events based on whether or not they accepted climate change' (2015: 15). The same basic pattern of weather can be used to support one view or the opposite view through cherry picking of data and the (often implicit) comparison data (*colder* and *warmer* are, after all, comparative terms).

Thirdly, Ross argues that the false consensus bias is a form of ego-defence. We protect ourselves (and justify our actions and beliefs) by overestimating how common our own behaviours and attitudes are. Behaviours different from our own are perceived as 'uncommon, deviant and inappropriate' and very revealing about underlying disposition. You can see how this can contribute to the *conflict* over climate change. We are not good intuitive psychologists. We do not sample data; we mix with people like ourselves who share our views of the world. We end up with a false consensus. We're not just right, we constitute the moral majority; we are the norm; there is nothing strange or distinctive about us. Only those who disagree with us are strange and distinctive.

You can see how this can operate in the case of trophy hunting, which has its own sub-culture of publications, websites, blogs, organised trips (and the dedicated infrastructures to support them) and conventions which allow participants to mix with people like themselves, who share their values and beliefs and foster the false consensus effect: 'Everybody knows that it's right to hunt and kill animals, that's just the natural way. Why shouldn't we be proud of it?!'

The Safari Club International Annual Hunters' Convention is one of the major gatherings for trophy hunters. It was held in Reno, Nevada in January 2019. The hunters talked about the thrill of the chase, the excitement of the kill, their pride, their momentary sadness after the kill (sometimes), their closeness to nature (which comes at a significant price), their 'realistic' perspective on the reality of life and evolution, their realism and closeness to nature, their appreciation of beauty in the animal kingdom, their activities, their wealth, their courage, their abilities, their know-how, their cunning. There is a real 'buzz' at the convention: 'We test ourselves against nature, we get to know ourselves, we love these animals that we kill in this epic battle. We love these animals [that we kill] better than anyone on the planet.'

The financial costs associated with trophy hunting are staggering; the upper limits are astronomical. In 2015, at auction, the Namibian government sold a hunting permit for the critically endangered male black rhino (*Diceros bicornis*) for US$350,000. According to Di Minin and colleagues from the University of Helsinki (2016), this male rhino was considered 'surplus' to the national black rhino management plan, but this did not stop the moral outrage.

But this is the buzz of the Safari Club convention—it's like Formula 1, all that glamour and glitz, but with guns. And so real, so close to nature, high finance, no barriers, countries and governments bending over backwards with their 'surplus' stock to give you what you can afford, power, influence, control, admiration. This is the norm, if you're there, if you're one of them (or one of us); a culture of (a curious kind of) love and manly values (interestingly even with the female hunters), the paradox of killing the thing you love, the insight, the understanding, the depth of it all—almost spiritual—life is paradoxical/life is death/death is life.

But outside this culture, from a little further away, things may look very different. An article headed 'Creepiest Festival for Trophy Hunters Is Kicking Off This Week' (*The Dodo*, 10 Jan. 2019), reads:

> Among the bustling crowd, a trio of lions with snarling teeth stares down at shoppers as they pass by. With eyes glazed over and manes perfectly manicured, the big cats just sit there frozen like statues. Under

> the bright lights, their huge teeth and giant whiskers shine. Although magnificent, these lions aren't alive. They were shot dead by a trophy hunter and now they're stuffed and up for sale at a creepy convention with dozens of other dead animals who were killed the same way.

At the convention, attendees can buy guns, gear, mementoes, taxidermised animals, animal pelts, animal body parts including heads and teeth, but most important of all, safari trips (e.g., a nine-day four-trophy package with a special offer: Buy this today and we will buy your plane ticket!).

In 2016, based on federal tax returns, the Safari Club International is said to have earned more than $7.7 million by hosting the convention. The club is a popular and booming business which normalises the behaviour and the feelings of achievement from killing an animal, regardless of how it is accomplished (and how unfairly the odds are stacked). The organisation and its members evidently see no psychopathology in any aspects of trophy hunts. However, once we start using terms like 'creepy' as in the online piece in *The Dodo*, we explicitly suggest the unusual nature of the activity, with connotations that it is 'obnoxious' and 'weird' and hint at the 'psychopathology' driving the business and the activity.

The false consensus effect is observed in various domains including climate change as we have just noted, climate change believers and deniers over-estimate the commonness of what they believe and do, see their views as normal and appropriate given the available evidence, and make more confident and extreme assumptions about the underlying characteristics of the opposing group (see Beattie and McGuire 2018). But rarely are the attributions about underlying dispositions so confident, intense and emotionally charged as they are in the case of trophy hunting. The hunters are sometimes branded as 'psychopathic' or 'sick' by those who are opposed to hunting. Non-hunters are deemed 'weak' or 'ignorant' by those who hunt. The false consensus effect, as a form of bias in everyday social thinking, does not, of course, rely on *any* scientific evidence about the actual psychological characteristics associated with the activity in question. Whether there are any personality correlates of trophy hunting or similar activities is an interesting and important question that we will consider later.

So, what does the evidence tell us about the psychological aspects of trophy hunting? How may we go about answering the question about what drives it? The first point to make is that any attempts to answer these broad questions will necessarily involve consideration of human behaviour at several levels. There are published articles in the literature with titles like 'Why men trophy hunt', which might appear, at least based on the title, to offer quasi-definitive answers to the question, but they then restrict their view to one (and only one) level of description and explanation. In the case of this article published in 2017 by Chris Darimont and his colleagues, the argument is restricted entirely to the evolutionary level (see Darimont, Codding and Hakes 2017).

Construing trophy hunting in primarily evolutionary terms and viewing trophy hunting in the context of the activities and practices of both contemporary hunter–gatherer societies (such as the Meriam people of Australia who hunt green turtles or the Maasai of Eastern Africa who hunt lions) and the activities and practices of our evolutionary forebears is very significant and illuminating but it cannot be the whole story. It may provide an important context for helping us understand the behaviour in question but may have less to say about why only certain members of our contemporary societies are drawn to it and why others are repelled by it.

Evolutionary models in this domain also put their emphasis on *men* as hunters of big game (that's how hunting of large game is organised in contemporary hunter–gatherer societies), but when it comes to modern-day trophy hunting, women may be a minority (in statistical terms) but they are a very significant minority, especially in terms of the propagation and advertising of the activity. Consider, for example, Kendall Jones, student, cheerleader, hunter. She is the pin-up girl of trophy hunting, treated as a celebrity by her thousands of fans and followers on Facebook, and lauded by the hunting lobby. Her interview with Bill McGrath, Legislative Counsel of Safari Club International (SCI), posted in 'First for Hunters' in 2014 reads: 'She is a seasoned hunter that has been lucky enough to travel to Africa to hunt the Big Five; work alongside organisations to provide meat to local communities and help treat wounded animals.'

The text accompanies a beaming and very photogenic Kendall sitting behind a large lion that she's just killed with a crossbow. She wears a lot of make-up in the African bush, her left arm sits nonchalantly on the animal's rear. The crossbow is just in front of the dead animal. She is beaming for the camera. The writer of the piece sees no obvious irony in juxtaposing the comment that she likes to 'help treat wounded animals' with the fact that she evidently *loves* killing fit and healthy animals, as if these two statements can somehow fit together without further explanation or justification.

Darimont and his colleagues (2017) noted that, 'Given that women in hunter–gatherer societies overwhelmingly target small, predictable prey compared with men (Codding et al. 2011), there are now seemingly puzzling examples of female trophy hunters.' Evolutionary biologists have little to say about these *puzzling* examples of female behaviour when it comes to both the killing of large prey and the resulting visual representations of trophy hunting even though both are important facets of the phenomenon. Some females are *ardent* trophy hunters and given Kendall Jones's visibility and presence on social media, this visual (indeed iconic) representation of the attractive female trophy hunter with the dead animal at her feet is a very important element in the spread of its popularity. To borrow a phrase from a few years ago, it iconically represents 'girl power' and exhibits power, authority and dominance. However, the Safari Club interviewer was keen to note that Kendall additionally 'works with organisations to help treat wounded animals.' The interviewer wants us to see that Kendall possesses the explicit 'feminine' characteristic of caring for animals to counterbalance the implicit 'masculine' characteristic of wishing to subjugate and kill large prey. The psychological process of depicting others (or ourselves) using language showing we care about animals and at the same time are capable of murdering them for sport—and proud of it—will also be a core consideration for any attempt at an explanation. Examples of this constructive process will be introduced from domains other than trophy hunting to show how this process can operate more generally.

For example, in their accounts of their own actions, terrorists will often construct themselves as kind-hearted and caring individuals whilst at the same time describing the barbaric acts in which they engage (Beattie 2004). The discourse of terrorists will be considered here not because *any* direct comparison is being made with trophy hunters per se; rather I consider terrorists because they also engage in activities that morally outrage many people. Their constructions of their positive qualities are often paramount when they talk about their actions performed in pursuit of some political or ideological goal, as is their (explicit or implicit) blaming of the victims (Beattie 1992: 2004). And so too with trophy hunters, as we shall see later.

To understand trophy hunting, we need a broader perspective on human psychology and the processes that maintain it than the view offered by the evolutionary perspective alone. We need to consider features of individuals like gender or aspects of personality to consider why some people are drawn to trophy hunting and others are not. To discuss individual differences and psychological needs, we must understand how trophy hunting operates in a social and interpersonal context, and more specifically how both the processes and results of trophy hunts are represented and displayed for other members of the group and for society more generally.

This necessarily involves consideration of the automatic and non-conscious aspects of behaviour like certain types of nonverbal communications (such as types of smiles) when hunters are photographed with their 'harvest' (see Beattie 2016). We must also examine the more deliberative and conscious aspects of the hunters' images (what is included and what is not, positioning, appearance of the dead animal and other factors). Thus, any attempt to understand the psychology of trophy hunting necessarily involves analysis of the semiotics of the activity and the roles of different systems of cognition and communication (Kahneman 2011; Beattie 2018a).

An understanding of the semiotics and communicational value of trophy hunting may give us some insight into the psychological needs satisfied by this activity and this insight may act as an important bridge to those personality theorists who have attempted to see whether any unique personality characteristics (or clusters of personality dimensions) are linked to trophy hunting. Personality, aspects of interpersonal communication and social signalling are intimately linked in everyday life and are relevant to understanding these aspects of trophy hunting.

Of course, we must also consider the language and discourse of trophy hunters and their supporters (as noted earlier) as they describe, explain and justify trophy hunting or deliberately claim that justification is not necessary. In Kendall Jones's interview with the Safari Club, when asked how she justifies trophy hunting, she stated, 'I have a hard time with the word "justify" because it implies that I am doing something wrong and have to make excuses for my actions'. At times, we need a consideration and analysis of the functions of language through a process of discourse analysis (Potter and Wetherell 1987; Edwards and Potter 1992; Beattie 2018a) as a way of determining how trophy hunters give meaning to their activities in terms of their routine linguistic constructions of their actions. We also need a functional analysis to determine why this is and why they use certain linguistic constructions. Finally, some research in neuroscience is required to demonstrate how hunting activity may influence the associative connections and basic architecture of the human brain.

This book thus attempts to address the psychology of trophy hunting at several levels, from an evolutionary perspective to provide some background about the evolutionary pressures on hunting and how it operates, through communication and semiotic theory to understand the signalling value of trophy hunting, with emphasis on nonverbal communication as a way to generate fresh insights into psychological states, in terms of personality theory to highlight and explain individual differences, through discourse analysis to understand how meanings are constructed and contested in this domain, and using basic neuroscience to understand the neuroscientific corollaries of all these results. This seems to be the first time that a multi-level approach has been used in this way and, by transcending traditional discipline boundaries, it aims to offer new conceptual and theoretical connections between different approaches to this significant and emotionally charged topic and hopefully raise important issues for both debate and, perhaps most critically, for future research in this area.

SUMMARY

1. There has been a great deal of highly charged emotional debate and moral outrage about trophy hunting over the past few years (reaching a peak after the killing of Cecil the lion in 2015), with much argument about the ethics of trophy hunting and whether, even if the fees from trophy hunting could (or do) feed into the local communities, it could ever be justified.

2. Clearly, the acts of paying enormous sums of money to travel to Africa to kill certain trophy species, particularly the Big Five for 'sport' and then displaying the carcasses in images which include smiling hunters and the means of killing evoke powerful emotions.
3. In this book, we will explore the psychology of the trophy hunter in a critical fashion, considering a wide range of different psychological perspectives.
4. Judgments about the ethics of trophy hunting are often bound to judgments of character and the motivations of the proponents.
5. The ethical perspective of 'consequentialism' which is that the ends justify the means—in this case that trophy hunting helps conservation—is not considered an adequate model for ethical judgment in a number of domains.
6. Arguments against trophy hunting are often dismissed as emotional rather than rational. Research in neuroscience, however, suggests that emotional responses often precede and assist rational decision-making in adaptive ways. In other words, emotional responses are not necessarily maladaptive or entirely irrational.
7. There is a great divide between proponents of trophy hunting and those opposed to it, often leading to conflict between the two groups, which we may try to understand by considering how people normalise their behaviours through everyday interactions and the implications of this for our psychological reasoning about our own group and the 'other'.
8. Female trophy hunters appear to fit uncomfortably with the evolutionary theory, often evoked by trophy hunters, that big game trophy hunting is a natural human activity, fixed by our evolutionary heritage.
9. Understanding trophy hunting will necessarily involve analysis of various systems of human cognition, including fast automatic (and nonverbal) processes in the depiction of the kill in social media images, along with slower, more reflective processes observed in talk and elsewhere, as the social meanings of the acts are described and justified.
10. Trophy hunting necessarily requires a broad and multi-levelled psychological perspective if we are ever to attempt to fully understand it.

Chapter 2

An evolutionary perspective

The hunter and his entourage with a dead rhinoceros.

For us hunting wasn't a sport. It was a way to be intimate with nature, that intimacy providing us with wild unprocessed food free from pesticides and hormones and with the bonus of having been produced without the addition of great quantities of fossil fuel. In addition, hunting provided us with an ever scarcer relationship in a world of cities, factory farms, and agribusiness, direct responsibility for taking the lives that sustained us. Lives that even vegans indirectly take as the growing and harvesting of organic produce

kills deer, birds, snakes, rodents, and insects. We lived close to the animals we ate. We knew their habits and that knowledge deepened our thanks to them and the land that made them.

Ted Kerasote
Merle's Door: Lessons from a Freethinking Dog

Ted Kerasote is a journalist, editor and naturalist who has written at length about the connections between human beings and the natural world. He grew up in Manhattan but wrote that he never felt that he really belonged there. Instead he discovered himself in his holiday home in Oyster Bay in rural Long Island. There, he 'became a native... on the water, fishing and hunting'. Hunting in his view brings people back to their natural heritage. He is a critic of environmental activism which he says, 'despite the very best of intentions, disrupts the natural environment and brings suffering to animals'. Kerasote isn't writing about trophy hunting but hunting more generally and the way that this can bond humans to nature. Therefore, it is 'natural', both in the present and throughout our evolutionary history. But how does trophy hunting fit into this evolutionary argument?

In an article published in *Biology Letters* in 2017, Chris Darimont and his colleagues commented that although trophy hunting has been analysed in the past in terms of the motivations behind trophy hunting, and the 'multiple satisfactions' that hunters seek from trophy hunting (principally affiliation, appreciation and achievement identified using content analysis of hunters' blogs), they point out, rather surprisingly, that no evolutionary explanation had been attempted before. They suggest that if we want to understand trophy hunting in the developed world from an evolutionary perspective, we should focus first on extant hunter–gatherer societies where large prey is often targeted. They start with the most basic of questions: why do hunter–gatherers target large prey in societies that we can observe today? From this they hope to be able to understand the characteristics and features of the hunting of large game in evolutionary terms.

This question may seem straightforward enough and some evolutionary anthropologists provided the most direct (and obvious) of all answers: large prey are targeted because of their obvious nutritional benefits to the hunters and their immediate families. But other evolutionary anthropologists have argued that this is simply not the case. Large prey provides too much food at any one time. The food must be shared with other non-immediate family members, which does not fit so readily with evolutionary theory where survival of one's own genes through close biologically-related family members is the primary (or only) consideration (Dawkins 1976). Therefore, the argument goes, the sharing of this food with non-immediate family members must be evolutionarily advantageous in *some* way (Kaplan 1983).

What exactly are the advantages of sharing this food with non-relatives? Again, the most obvious answer is that sharing meat is in the interest of the hunter because it results in future reciprocation or some other form of pay-back (Winterhalter 1986). Wiessner (2002) reviews the empirical evidence for this and shows that the reciprocation of food is rarely proportionate to the amounts given, and in many cases, food is not directed to the original giver by way of repayment. Wiessner does acknowledge that some delayed reciprocation may occur at times of need for the original hunters, but she stresses that this is unlikely to be the full story (Gurven et al. 2000).

Kristen Hawkes (1991, 1993) offers a different explanation. She evokes 'costly signalling theory' to provide a different perspective on this issue. The basic premise behind this theory is that certain animals (including humans) use conspicuous display as a form of communication that signals inclusive fitness, thereby making the animals more successful in mating and propagating their genes. However, these displays come at a cost in that they require a considerable amount of effort, risk, economic resources and time to work (see Griskevicius et al. 2007).

Take, for example, the peacock that displays its tail to attract attention during courtship in order to signal the quality of its genetic makeup by the sheer elegance and spread of its feathers. This is obviously a costly signal in that the elaborate display makes the peacock more vulnerable to predators. This theory proposes that hunters target large game to attract social attention—to increase their social standing in the group and to make them more successful competitors for mates and for group support. Large game hunting is an effective form of showing off, not unlike the peacock's tail spread. It is a form of costly display—*costly* because it is inherently very dangerous. Wiessner (2002: 409) argued that:

> Procurement and sharing of large game as a collective good provides an honest signal about the hunter's strength, skill, risk taking, and leadership, which is costly in ways not subject to reciprocation. The value to the recipient is the possibility of evaluating the signaller's abilities, qualities, or motivations by attending to the signal rather than discovering these attributes by costlier means. The value to the signaller resides in the information conveyed to those who receive the signal (Zahavi and Zahavi 1997) with resultant increased likelihood of being chosen as a mate or ally, or, in particular, being deferred to in status competition through which individuals may gain access to mates and resources.

In the case of large game hunting, members of the group are attracted by the display of the kill because they gain, not just from the meat, but *information* from and about the hunters and the hunt itself. The audience then uses this information, in the words of Hawkes and Bird (2002: 61) 'to their own advantage in the numerous decisions of social life'. This theoretical idea considers the carcass of the dead animal 'not only as food but also a signal of the costs associated with the hunter's accomplishment' (Darimont et al. 2017: 1). The carcass signals social status and prestige; it is a symbol and (very importantly) a reminder that the individual hunter possesses the skills, knowledge and attributes to kill such dangerous animals.

Darimont reviews the evidence from contemporary hunter–gatherer societies to determine whether this idea has any merit. For example, the men of the Meriam people of Australia hunt green turtles (*Chelonia mydas*) at sea in dangerous waters and the Maasai of Eastern Africa kill lions. Successful hunts 'signal underlying qualities to rivals and potential allies....Meriam turtle hunters, who gain social recognition, get married earlier to higher-quality mates, and have more surviving children' (Darimont et al. 2017: 2).

Darimont and his colleagues argue that costly signalling theory can be applied directly to trophy hunting and claim that the activity is not about the consumption of meat, but primarily about the display of achievement and status in the thousands of photographs of the hunters and their prey. But the interesting question then becomes what specific qualities do these displays reflect? Given that critics of trophy hunting argue that the hunts are unfair, rigged competitions, with the animals in confined areas (of varying dimensions) facing hunters with powerful assault rifles (who also use expert trackers and guides who may well themselves have the skills and knowledge envisaged by costly signalling theory), what actual dangers do Western trophy hunters face?

There is *some* danger, of course. Pero Jenic, a trophy hunter from Croatia, was killed in January 2018 as he hunted lions bred in captivity. He was, however, killed by a stray bullet rather than by a lion, as he tracked the lions near Setlagole in the North West Province of South Africa. If a hunter faces little danger (relatively speaking) and requires potentially little skill, how can costly signalling theory be relevant to the hunt? Darimont says that there is a fundamental shift from cost in terms of danger or difficulty to one of financial cost although many hunters want to maintain both aspects, as we shall see. In other words, money replaces the laws of the jungle. But this is not the first time that money and consumption have been deemed relevant to evolutionary theory.

Conspicuous consumption: Trophy animals as commodities

The public display of status through purchased goods has been defined as 'conspicuous consumption.' The economist Thorstein Veblen first coined the term in 1899 in his classic book titled *The Theory of the Leisure Class*. He used the term to define 'the advertisement of one's income and wealth through lavish spending on visible items' (Heffetz 2011: 1101).

Here consumption is understood as a communicational act that occurs in a social context and that interlocutors can interpret. The main theoretical perspective on conspicuous consumption is premised on the costly signalling theory we have just considered. For an action to qualify as costly signaling, it needs to meet four criteria: (1) It must be costly to the signaller in terms of economic resources, time, energy, risk or some other significant domain. (2) It must be easily observable by others. (3) The display must ultimately increase the odds that the signaler will gain some fitness advantage such as increased ability to attract desirable mates. (4) The signal must serve as an indicator to potential mates of some important trait or characteristic, such as access to resources, pro-social orientation, courage, health, or intelligence (Griskevicius et al. 2007: 86). Note that *access to resources* is included in this list alongside courage, health and intelligence.

Expensive or luxury purchases (Veblen 1899) obviously meet these criteria (and commercial advertising, of course, is based upon this fundamental idea). Veblen was not concerned with trophy hunting, but it is easy to conceptualise dead trophy animals, particularly, the Big Five as expensive *purchases* with high signalling value. Conspicuous consumption may explain why people covet luxury goods or have the desire to hunt rare (and expensive) animals. The reason may simply be that the ostentatious purchases of luxury goods or the killing of rare species (when it involves considerable cost) will attract more friends and sexual partners through our ability to signal that we have access to the appropriate financial resources (Black and Morton 2015).

However, there are additional considerations here. Some suggest that costly signalling theory does not always lead to negative conclusions with respect to consumption and the environment (more and more expensive purchases) and that it may also apply to the opposite end of the environmental spectrum and produce *positive* environmental behaviours (see Beattie and McGuire 2018). For example, Griskevicius et al. (2010: 392) argued that people are indeed willing to act *pro-environmentally* at times because it enhances their social status. They cited the example of the Toyota Prius (a 'green' hybrid car that costs more than a conventional equivalent) and compared it with the Honda Civic (a cheaper but highly efficient equivalent standard car). In a 2007 survey conducted among customers who purchased the Toyota Prius, advertised as 'the planet's favourite hybrid', over half of the people surveyed said that the main reason for buying the Prius was that it 'makes a statement about me.' Only a quarter of the customers bought the car because it generated lower emissions (Maynard 2007). One owner openly admitted 'I want people to know that I care for the environment.' In other words, the main reason for buying a Prius may be social identity and elevation of social status through consumer choice.

Griskevicius et al. (2010: 395) empirically investigated the connection between pro-environmental behaviour and elevated status. Participants in their study were given a 'motivational prime' in the form of a short story aimed to prime their motivation to attain high status on an unconscious level. The story required them to imagine that they were 'graduating from college, looking for a job, and deciding to go work for a large company because it offers the greatest chance of moving up'. The story went on to describe the upmarket place of work with its 'upscale lobby and nice furniture.' Near the end of the story, readers learned 'that they will have an opportunity to receive a desirable promotion. The story ends as the reader ponders moving up in status relative to his or her same-sex peers'. In a control condition, participants were also asked to read a story of a similar length that was not designed to prime social status. Instead, they 'read about losing a ticket to an upcoming concert and searching through the house. After the person finds the ticket, he or she heads off to the concert with a same-sex peer'.

The study also involved a second control condition where participants did not read a story, but simply had to make product choices. After the various manipulations, participants had to imagine that they were out shopping for three products: a car, a household cleaner, and a dishwasher. Each product had a luxury option and an environmentally-friendly option. Both options were similar in price, were made by the same manufacturer and had three key features. Participants saw the products on a computer screen in random order and were asked: 'If you were out shopping for a car, dishwasher or household cleaner, which of these two products would you buy?'

The study revealed that participants in the control condition were more likely to choose the luxury options than the pro-environmental options, whereas, in the experimental group, participants primed with the status motivation story were more likely to choose the pro-environmental option. The authors concluded that 'activating status motives led people to increase the likelihood of choosing pro-environmental green products over more luxurious non-green products' (Griskevicius et al. 2010: 396).

This study tells us that pro-environmental consumer choices can relate to social status and that it is possible to prime this form of behaviour. Griskevicius et al. then considered the effects of social context by investigating the choice of 'luxurious non-green products' and 'green products' in a private setting (shopping online) versus a public setting (shopping in a supermarket). Participants again read the same story designed to prime status motivation, with a control group reading a story unrelated to status motivation. For the private setting condition, participants were told to 'imagine that you are shopping online by yourself at home' and public setting participants were told to 'imagine that you are out shopping at a store.' Participants then had to indicate their preferences between three green versus three non-green products. The items were a backpack, batteries, and a table lamp. Again, each product had green and non-green alternatives that were similar in price and manufactured by the same company.

The results revealed that when participants in the priming condition had to imagine that they were shopping in public, they showed an increased preference for green products compared to the control condition. However, when shopping in the private condition, participants in the priming condition showed *decreased* preferences for green products. The authors concluded that:

> When purchases are being made in private—when reputational costs were not salient—activating status motives appears to somewhat increase the attractiveness of luxurious (non-green) products... status motives increased attractiveness of pro-environmental products specifically when people were shopping in public. When people were shopping in private, however, status motives increased desire for luxurious, self-indulgent non-green products (p. 397).

In other words, when people were aware that their choices could be observed by others and had the possibility of influencing other peoples' perceptions, they were more likely to choose pro-environmental products.

Griskevicius, Tybur and Van den Bergh (2010) also investigated what happens to behavioural choices when the green and non-green items are priced differently. They found that the experimental participants were more likely to choose green products when they were more expensive than non-green products. However, when the non-green products were more expensive and status motivation was activated, the green items were selected less often than their more expensive non-green counterparts. In other words, price can be more effective than environmental features in signalling status.

The research by Griskevicius and his colleagues suggests that costly signalling theory may be relevant to pro-environmental behavioural choices, particularly in the presence of others, when it becomes a public display. But the research still suggests that the display of wealth or financial resource can outweigh this. The results may well cast some light on trophy hunting because the display reflects the hunter's socio-economic standing and the qualities that underlie that socio-economic standing. However, if the hunters can draw upon the discourse and associated linguistic frames of (1) conservation and caring about the natural environment as evidence of pro-social orientation (2) the discourse of danger, courage, and hunters' knowledge, as envisaged in costly signalling theory as further public displays of admirable traits and characteristics, these factors may well augment the power and efficacy of trophy displays.

Language as a self-presentational medium can and does operate around physical and more primitive (and evolutionarily significant) physical symbols such as a dead animal on public view. We thus may observe a form of multi-modal communication involving both nonverbal and body signals operating alongside language for self-presentational purposes. But, of course, the language of the hunt, the *conservation* arguments, the description of the cunning and the skill of the hunt do not produce the same immediate results as the display of a dead animal. Blogs, articles, and contributions to online forums take time to write. That is why the immediate body language of the hunter posing with the dead animal is critical. The body language of the hunter along with the display and arrangement of the dead animal conveys an immediate and powerful nonverbal message (Kahneman 2011). We will consider aspects of this body language in Chapter 3.

Other issues related to trophy hunting need also to be considered. Displays of wealth and financial standing are viewed from a socio-biological perspective to be important aspects of attracting a mate by men on the (implicit) understanding that such displays signal that they have the financial resources to raise any offspring. From this view, trophy hunting is very similar to the displays of luxury goods, sports cars and boats. It is essentially a form of competitive signalling, and social media participation allows such competitive signalling to be broadcast to a huge audience. Of course, if there is such a compelling evolutionary reason why men should engage in this form of behaviour, evolutionary factors do not explain why so many men disavow it completely.

Is the disavowal simply the behaviour of men who lack the financial resources to compete? Is their rejection of this activity a form of defence mechanism that allows them to maintain their self-esteem? Why are some men drawn to trophy hunting rather than being content with a sports car or a yacht? And how does this concept apply to women? After all, from a socio-biological perspective, the signalling of resources is thought to be the domain of men who use it to attract mates. Women, it is argued, have other signalling prerogatives such as youth, beauty and fertility (Dawkins 1976).

The evolutionary perspective on trophy hunting and the concept of conspicuous consumption set this behaviour in an appropriate context but leave many core issues unanswered and for that reason we turn now to other lines of enquiry.

SUMMARY

1. In this chapter, we considered the evolutionary perspective used by many to argue that hunting large prey is a *natural* way of securing sufficient food for families and is hence a reflection of the natural order.
2. A careful review of the evidence, however, suggests that even in evolutionary terms, the killing of large prey was always more than a method of securing ample food sources for immediate families. It also served as a social signal to the group about *inclusive fitness.*
3. In the contemporary world, a dead animal is not necessarily a signal of courage, skill, or cunning—the usual signs of inclusive fitness, but a signal of material wealth.
4. The evolutionary argument leaves many questions unanswered, for example, why only certain men are drawn to trophy hunting rather than being content with a sports car or similar trophy signal, and why there is a significant proportion of *female* trophy hunters. From a socio-biological perspective, the signalling of resources is thought to be primarily the domain of men in the attraction of mates.

Chapter 3

Psychological motivations

Expressed and hidden

Carefully posed and very proud.

We have never understood why men mount the heads of animals and hang them up to look down on their conquerors. Possibly it feels good to these men to be superior to animals, but it does seem that if they were sure of it they would not have to prove it. Often a man who is afraid must constantly demonstrate his courage and, in the case of the hunter, must keep a tangible record of his courage. For ourselves, we have had mounted in a small hardwood plaque one perfect borrego [bighorn sheep] dropping. And where another man can say, "There was an animal, but because I am greater than he, he is dead and I am alive, and there is his head to prove it," we can say, "There was an animal, and for all we know

there still is and here is proof of it. He was very healthy when we last heard of him."

John Steinbeck
The Log from the Sea of Cortez

As well as being a very famous novelist, John Steinbeck wrote a number of works of non-fiction. *The Sea of Cortez: A Leisurely Journey of Travel and Research* is a non-fiction work written jointly by Steinbeck and Ed Ricketts, an American biologist. It details their journey around the Gulf of California (called the Sea of Cortez in the book because it sounded more glamorous). The book is both a travelogue and a biological record of the creatures they encountered, as well as musings by both authors about the place of humans in the environment. It also raises broader ecological concerns. Steinbeck apparently thought this was his best work and also said that he looked forward to the 'rage and contempt' of the critics at this rather surprising new direction in his work.

Steinbeck poses a fundamental question in the quote above: if people feel superior to other animals, why do they need to prove it with the heads of animals mounted on their walls? Steinbeck raised this fundamental psychological question in the wittiest of ways. So why do men and women trophy hunt, and what are their underlying motivations?

To find out why people pursue various activities, the usual practice employed by psychologists and everyone else is to ask them. You can choose to do this in an interview or via a questionnaire. Interviewing is quick and easy, but, of course, it assumes that people know why they do various things and have access to their underlying attitudes, beliefs and deep-seated motivations, and, in addition, are keen to present them to you as truthfully and as accurately as possible. This, of course, is not always the case.

The limitations of self-reports for describing behaviours

Psychologists working in some domains have found this to their cost. For example, during interviews with people about their attitudes towards sustainability and the environment, many report very positive attitudes that often do not align well with their *actual* behaviours (Beattie 2010). This may seem rather surprising given the (apparent) substantial body of research which suggests a significant and positive relationship between attitudes and behaviour when it comes to sustainability. Schlegelmilch et al. (1996: 51) reported that 'attitudes are the most consistent predictor of pro-environmental purchasing behaviour'. Honkanen et al. reported 'a significant relation between attitude and intention to consume organic food' (2006: 426). Dahm et al. reported that 'attitudes were significant predictors of consumption behaviours and practices....Positive attitudes toward organic foods and other environmentally-friendly practices significantly predicted similar behaviours' (2009: 195). Barber et al. reported 'a strong and significant relationship between attitude and willingness to purchase environmentally friendly wine' (2009: 69).

However, none of these studies examined actual behaviours. Rather, they focussed on *self-reports* of behaviour, intentions, or willingness to consume environmentally friendly products. Roy Baumeister and colleagues (2007: 396) have commented that although psychology may call itself the science of *behaviour* 'some psychological sub-disciplines have never directly studied behaviour'. They also noted that 'a remarkable amount of "behaviour" consists only of marks on self-report questionnaires. Sometimes these questionnaires ask participants to report what they have done, will do, or would have done. More often, they ask people to report what they think, how they feel, or why they do what they do' (p. 397). The responses to issues regarding the environment and climate change may also well be impacted by social desirability and reporting biases.

The relationship between *actual* environmental behaviour and self-reports of such behaviour is often much more problematic (Beattie and McGuire 2016). For example, Tsakiridou et al. (2008) explored the relationship between attitudes and behaviours towards organic products. They found that 50% of participants *reported* that they preferred to buy organic products. This finding, however, was contradicted by actual consumption data in that only a small proportion of those who expressed positive attitudes towards organic products purchased organic products.

Corral-Verdugo (1997: 135) randomly selected 100 families in Mexico who were required to report the amounts of glass, aluminium, newspapers and other materials they reused and recycled. These reports of behaviour were then compared with direct observations of reuse or recycled items. The researchers found that 'beliefs (assessed verbally) only predicted the self-reported conservation, while competencies (assessed nonverbally) were only related to observed behaviour'.

Similarly, Fielding et al. (2016: 90) measured self-reported household recycling and water conservation behaviour as well as actual recycling and water use. Their results showed a 'weak relationship between self-reported household recycling and objective measure of recycling' and a 'weak relationship between self-reported water conservation behaviour and objective household water use'.

Kormos and Gifford (2014: 360) performed a meta-analysis of the validity of self-report measures of pro-environmental behaviour and concluded that 'self-reports are only weakly associated with actual behaviour'. They identified some of the factors responsible for this weak relationship including the fact that self-report measures may be prone to exaggeration and, because self-report measures are subjective by nature, descriptors such as *often* may mean different things to different participants. In addition, self-reports of behaviour may 'reflect individuals' perceptions of their behaviour (Olson, 1981), behavioural intentions (Lee, 1993), or other—sometimes false—beliefs and attitudes (Rathje, 1989), rather than objective behaviour.' They also note that 'limited memory or knowledge may also reduce the accuracy of self-reports (e.g. see Warriner, McDougal and Claxon 1984)'. In other words, self-reports of our attitudes may not be good predictors of our actual behaviour and self-reports of our own behaviour may be heavily biased.

Or consider a different domain with equally important real-world consequences. When it comes to race and racial differences, few people in contemporary society would openly report that they are racially biased. However, regarding actual behaviours including judgments and decisions with real-world consequences such as shortlisting applicants for various posts, racial biases become immediately apparent (Beattie 2013). This may be more than just a reporting bias, arising from the fact that self-reports of racism or racist sentiments are no longer socially acceptable or socially tolerated in ways that they once were. In 1942, for example, only 35% of white Americans reported that they would have felt comfortable with black neighbours; by 1963 the figure was 64%. Also in 1942, only 42% of white Americans reported that it was acceptable for black passengers to board buses with them. By 1963, the figure was 78%. The dynamics may involve something much deeper.

It is possible that certain aspects of our everyday thinking and actions are not open to conscious introspection or available to conscious awareness. The Nobel Laureate and behavioural economist Daniel Kahneman (2011), for example, talks about *systems* of human cognition rather than *the* system of human cognition. He maintains that only one of these two systems (System 2) is conscious and reflective (as well as being slow and rational). System 1 involves much faster decision making and actions. Kahneman describes System 1 as an emotionally-based workaholic which operates in *non-conscious* modes.

In recent years, psychologists have developed and refined new ways of measuring these fast, non-conscious processes and implicit and unconscious attitudes. This measurement requires no conscious reflections on the part of the participants and no reporting of internal states. Some researchers indeed now claim there is no point in asking people about their attitudes because they are unaware of what sets up and establishes their 'predispositions to act' (the most common definition of *attitude*). The best known psychological method for measuring implicit attitudes is the Implicit Association Test (IAT), developed by Anthony Greenwald. It requires a participant to simply press a key as quickly as possible when he or she sees a face or a name and a word together on a screen. Greenwald found that white people are quicker at associating white faces and names with the concept *good* than they are at associating black faces and names with *good*. Similarly, white people are slower to associate white faces and names with the concept *bad* than associating black faces and names with *bad*.

In the first web-based experiment of its kind known as Project Implicit, 600,000 tests were carried out between October 1998 and December 2000. The results indicated that white participants tended to *explicitly* endorse preferences for white people compared to black people (a moderate pro-white explicit preference), but *implicitly* (in terms of IAT score) they demonstrated much stronger preferences for white names and faces (a *strong* pro-white *implicit* bias).

Conversely, black participants demonstrated *explicit* preferences for black people compared to white people and yet, remarkably (some would say) in the IAT, black participants demonstrated weak *implicit* preferences for white names and faces. In other words they were quicker to associate black names and faces with the concept *bad*.

In our research (Beattie 2013), we also found that white people are quicker to associate images of white faces with the concept *good* than associating non-white faces (black, Asian, Middle and Far Eastern) with *good*. They were also slower to associate white faces with the concept *bad* in comparison with non-white faces. In fact, we found that over 40% of the participants who explicitly espoused exactly neutral attitudes about race held very strong implicit attitudes which, in most cases favoured whites over non-whites (Beattie et al. 2013).

The IAT essentially measures associative connections in the human brain between two concepts: (perceived) racial background and the concepts of good and bad; based on the fact that individuals may well be unaware of these connections. It seems that *white* and *good* are more closely associated than *non-white* and *good* certainly for most white participants. Furthermore, this finding even applies to a proportion of non-white individuals who are plugged into these broader cultural stereotypes until they accept this association automatically and without reflection. This negative *implicit* attitude seems to be often at odds with the attitudes people express about race, and perhaps for the first time we can identify and measure some sort of inner conflict about race (with explicit and implicit attitudes at odds with each other) and determine the implications of this conflict for their actual behaviours.

We attempted to assess the importance of these unconscious implicit attitudes in one very significant judgment situation (Beattie 2013): the shortlisting of candidates for certain jobs, specifically university posts, and we did this for a very good reason. If we are ever to understand racial discrimination in job opportunities and social mobility that clearly exist, we must consider what factors influence the processes involved in selection. An article in *The Guardian* newspaper in 2011, when our research was beginning, highlighted the issue of possible racial bias at liberal meritocratic universities.

The journalist wrote that '14,000 British professors—but only 50 are black' and continued, 'Leading black academics are calling for an urgent culture change at UK universities as figures show there are just 50 black professors out of more than 14,000, and the number has barely changed in 8 years, according to data from the Higher Education Statistics Agency.'

At the time, only the University of Birmingham had more than 2 black British professors and only 6 of 133 universities employed more than 2 black professors from the UK or abroad. The article produced a great deal of hand wringing about racial bias in universities and what could be done to rectify it. The article included a picture of Harry Goulbourne, a black professor of sociology at London's South Bank University, who was quoted as saying that 'Universities are still riddled with *passive racism*.' The chief executive of Universities UK, an organisation of vice chancellors of British universities, was also quoted in the article: 'We recognise that there is a serious issue about lack of black representation among senior staff in universities, though this is not a problem affecting universities alone, but one affecting societies as a whole.'

Black representation in the academic world appears to be an international problem, with a similar picture emerging from the US, where only 5.4% of all full-time academic staff members at universities come from black and minority ethnic backgrounds (US Department of Education 2007). According to the *Journal of Blacks in Higher Education*, 'If we project into the future on a straight-line basis the progress of Blacks into faculty ranks over the past 26 years, we find that Blacks in faculty ranks will not reach parity with the Black percentage of the overall American workforce for another 140 years' (cited in Leathwood et al. 2009).

In our research, we presented the participants with curricula vitae (CVs) of either white or non-white job candidates. They were not told that the apparent race of the candidates had been changed while all other data remained the same. We wanted to see which candidates were shortlisted. Before the participants reached their final decisions, we used a remote eye tracker to see where exactly they looked at the CVs on the screens in front of them. We also measured their implicit and self-reported attitudes. The results were astounding.

White participants (mostly young, liberal-minded students and staff from a UK university who espoused no negative racial attitudes whatsoever) were ten times more likely to shortlist two white candidates for a lectureship post than two non-white candidates with exactly the same CV qualifications. The *implicit* attitude of white participants significantly predicted the ethnicities of candidates who were shortlisted for the post. The IAT scores of those white participants who shortlisted two white candidates (instead of one white and one non-white or two non-white candidates) for the lectureship post indicated a *strong* pro-white implicit bias. Self-reported attitudes towards different ethnic groups, on the other hand, did not have a bearing on the behavioural preferences for shortlisting white candidates over non-white candidates (Beattie et al. 2013).

The races of the participants also affected what sections they looked at on the CVs of white and non-white candidates in the minute before they had to make their shortlisting decisions. White participants spent more time looking at good information on the CVs of white candidates and at the bad information on the CVs of non-white candidates. Non-white participants spent slightly more time looking at the bad information rather than the good information of white candidates. Non-white participants looked equally at the good and bad information of non-white candidates. Moreover, the stronger the pro-white biases of the participants, as measured by the IAT, the more likely they were to fixate on the negative parts of the CVs of non-white candidates as they made their decisions.

Decision makers may think that they are making rational decisions when shortlisting for university posts, but their gaze fixations as they review CVs are affected by both their own race and the races of the candidates being evaluated. Participants with high IAT scores (strong implicit pro-white bias) spent more time looking at the good information on the CVs of white candidates compared with participants with lower IAT scores (less strong pro-white implicit bias).

In other words, our implicit (and largely unconscious) attitudes towards people from different racial backgrounds seem to direct our unconscious eye movements when we consider their CVs. Our *rational* decisions about the suitabilities of candidates may well be based on this biased pattern of gaze fixation. Gordon Allport (1935), the founder of social psychology, may have described the way that inner conflicts regarding racial prejudice were dealt with cognitively nearly 90 years ago. However, newer research in this area gives us some insight into the underlying mechanisms, including how unconscious attitudes direct eye gaze, that impact on decision-making behaviour. We may think that we are making rational decisions about a candidate's suitability for certain posts, but these rational processes and the way we scrutinise CVs are, in reality, dominated by largely unconscious processes.

This I suppose, is the reality of prejudice in action that works away below the level of consciousness. Unconscious attitudes influence unconscious behaviours as we gather the information the conscious mind needs in order to appear rational, orderly and fair. This is what Allport's pioneering research missed.

This research on prejudice and how it operates raises fundamental issues about the effectiveness of merely asking people to report their motivations to engage in any form of behaviour (such as candidate selection, sustainable behaviour or trophy hunting), and then accepting what they say as the absolute psychological truth. We may think we know why we do certain things, but our behaviour is often puzzling when it is more carefully examined. Why do we recycle less than we should and go on many foreign holidays despite reporting that we care deeply about climate change? Why do we only shortlist white candidates from an ethnically diverse set of applicants when we sit on a university appointments panel despite reporting that we care deeply about racial equality and the importance of diversity? Why do we kill animals that we say we love?

Any attempts to understand these contradictions may well involve a consideration of deeper, more unconscious psychological processes, and possibly the interactions of conscious and unconscious processes that together can achieve certain ends. One example of this type of interaction is the pattern of gaze fixation as we review CVs. The fixations are guided by unconscious implicit attitudes that identify weak points in the CVs of black and minority ethnic (BME) candidates and then they guide our conscious 'rational' decisions to keep them off the shortlists. Clearly, we need to be cautious about accepting verbal accounts at face value as indicative of underlying psychological motivations, attitudes or beliefs (without, at least, some degree of reflection and analysis).

Notwithstanding new research identifying and measuring implicit attitudes and unconscious processes in human decision-making and actions, we do, however, still need to listen to and analyse what people say for one very important reason: language is how we give meaning to what we do, both for others and for ourselves. For that reason, we need to analyse how language works to determine its function and how it is used to construct an individual's version of an event. But this analysis is not necessarily that straightforward.

When we talk or write about our experiences and our lives, we are keen (often subconsciously) to present ourselves in various ways (often in a good light), and build a coherent picture of our social world. Our descriptions of our acts and associated motivations may be shaped to provide a coherent narrative. This presentation of self does not always have to be so obviously deliberative and carefully and consciously planned. It may, on occasion, be routine and automatic. Presentation of self involves what we do, day in day out, sometimes without much conscious reflection. When we analyse spoken or written language, we need to consider carefully how this language is used and what it is doing (Potter and Wetherell 1987).

Based on interviews with trophy hunters and analyses of posts on hunting websites, some researchers have argued that a number of apparently distinct psychological motivations for hunting emerge in these accounts, specifically concerning the satisfactions that hunters derive from the activity (Ebeling-Schuld and Darimont 2017). The main satisfactions reported are achievement (feelings of satisfaction related to performance), appreciation (enjoyment of the experience) and affiliation (strengthening of personal relationships or enjoyment of the company of others).

Analysis of hunting stories: A critique

Ebeling-Schuld and Darimont (2017) analysed 455 hunting stories from online hunting forums from three regions (Texas, British Columbia and

North America), and coded 2864 individual phrases from these stories. The phrases served as the fundamental units of analysis for their study. They varied significantly in size and grammatical complexity—from single words to complete clauses. These phrases were then assigned to various categories.

You could, of course, argue that if you want to understand the satisfactions based on personal accounts, you might need to proceed differently and analyse the whole discourse to understand what is said and use the interpretation of individual linguistic components including phrases as parts of the analysis. In other words, you might use a top-down method rather than the bottom-up approach used here.

Ebeling-Schuld and Darimont provided examples of how they coded the text (see table below). First, they identified key *phrases* that varied in size from individual words like 'harvest' on line 4 to a full clause as on line 7. The phrases were then assigned thematic *codes* such as 'thank you'; 'friends' and 'harvest'. The codes were then assigned to one of three satisfaction categories. Some codings stand out. It is odd that a single mention of 'harvest' on line 4, where 'harvest' is a core descriptor for a specific type of hunting activity (with a slightly more positive connotation than 'killing'), is automatically coded as an additional score on the achievement satisfaction category.

You might query the interpretation and coding above. Why does this scheme necessarily tell us anything about the main satisfaction derived from hunting? Of course, if you read this example as a whole, you can see that it is mainly about achievement (a combination of lines 1 and 4) or, more accurately, the hunter's boast about what he achieved by killing a beautiful buck with the help of his friends. The fact that the buck has certain desirable properties (the *beautiful* characteristic coded as *appreciation*) makes it more of an *achievement*. The cooperation among group members (*affiliation*) adds to the achievement. The group needed to co-operate to kill and harvest such a desirable animal. Perhaps, a top-down approach (reading the text as a whole) gives us more of a feeling about how the various elements work together. Language, after all, is a hierarchical and coherent medium of communication.

Number	*Key phrases*	*Satisfaction category*
1	Was really pleased with this buck	Achievement
2	Very thankful for all help from good friends	Affiliation
3	to	–
4	harvest	Achievement
5	this beautiful animal	Appreciation
6	I think my	–
7	eyes are being reopened to how beautiful this province is	Appreciation

Source: Ebeling-Schuld, A.M. and Darimont C.T. 2017. *Wildlife Society Bulletin*, 41, 523–529.

Nevertheless, the authors using this bottom-up approach reported that achievement was the dominant satisfaction in 81% of the ungulate (hoofed animal) and 86% of the carnivore stories. Appreciation was found to be nearly absent as a *dominant* (rather than subsidiary) satisfaction in the stories of hunting carnivores. The authors also found that both appreciation and affiliation played secondary satisfaction roles even when achievement was dominant. They noted that:

> The more pronounced prominence of achievement we observed in carnivore stories, coupled with fewer mentions of other satisfactions, provides important insight into the differences between hunters (or hunts) that target these different taxa. Greater difficulty and price associated with carnivore hunting may explain why achievement of a successful kill (or disappointment in a failed hunt) may be relatively more important than intrinsic (nonmaterial; i.e., appreciation and affiliation) satisfactions for carnivore hunters (p. 527).

Ebeling-Schuld and Darimont describe a clear hierarchy of reported satisfactions with achievement at the top, especially for carnivore hunting, at least in terms of online hunting forums, where 'tensions and stigmas may be removed and opinions can be expressed with fewer inhibitions' (p. 524). They suggest that these kinds of analyses could 'offer significant insight into wildlife management...designing regulations that focus on achievement satisfactions would be particularly successful (p. 528). In other words, we should design environments in a way that leads to more feelings of achievement.

However, a number of issues should be considered here. The first, as I have said, is the authors' basic approach to the analysis of the text using a bottom-up approach to identify and categorise key phrases as the fundamental units of analysis. This approach may not provide sufficient insight into how these basic components fit together. In 'harvesting' animals, the hunter, the hunted, the situation and the feelings are the basic components of the experience. Accounts are built around these components, but the meaning of the account as a whole is more than the meanings of the individual components. That's how language works.

There are then the issues of scoring the strength, intensity, and importance of the satisfactions derived from hunting as measures that may provide some insight into the fundamental motivations of hunters and their reasons for engaging in the activity by scoring the numbers of lexical items and phrases associated with each of these categories. The method at best provides a very crude index that may be misleading. Everyday experience teaches us that we sometimes mention events or feelings frequently to camouflage our real motivations. Sometimes we over-elaborate certain things at the expense of others. Quantity, as we all know, is not a sure sign of quality, intensity or importance of an experience. The achievement of hunting is killing an animal. The description is softened by choosing 'harvesting' as a more natural and seasonal expression, but the nature of the activity remains dominant. Are the more positive 'love' aspects (appreciation and affiliation) critical components of the basic motivation? Or, on the basis of quantitative analysis, are they not actually critical and merely represent additions to justify or explain the killing in the first place? Are appreciation and affiliation central, and do they cause hunters to deal with their 'conflicted minds' when it comes to hunting? (Beattie 2018a).

Others have taken a slightly different approach. In an analysis of 23 feature articles in 8 popular hunting magazines, Kelly and Rule (2013) suggest that the majority of articles they considered contained the 'love' and 'kill' words, revealing a multi-layered discourse. Their analysis suggests that the positive emotions go beyond mere appreciation and affiliation and at times approach a state they term 'love' and express regret at the killing ('You don't kill a magnificent animal like this without feeling a tug of regret pulling through the thrill', Braendel 2009: 153). Their conclusion is that love-and-kill themes exist together in over half of our sample, suggesting that the existing literature is flawed in framing loving and killing as separate constructs.

Unless we can acknowledge that conflicting values coexist in hunting, we cannot understand the wide array of perspectives that shape the complex interactions between humans and animals (Kelly and Rule 2013: 200). Understanding these conflicted emotions may be critical (if indeed they are conflicted, they may just be parts of a rhetorical or language-based strategy on the parts of the respondents). The attempt by Ebeling-Schuld and Darimont (2017) to pull such constructs (with concomitant emotional states) apart by scoring minimal linguistic units separately may not be the most effective approach. We need to understand how hunters give meaning to what they do through a coherent analysis of the text, and how it is constructed and used. A categorisation of the individual units may not give us the insight we need. It's like trying to understand the architectural form and function of a building by coding variations in the colours of the individual bricks used in its construction.

Online hunting forums will attract people who want to boast about their achievements. That's presumably why they've gone online in the first place. Is it any wonder that achievement emerged as the predominant satisfaction? How robust is this finding? It may well be robust. Most hunters want to boast about their achievements. This is plausible based on our considerations of costly signalling theory and the personalities of hunters (we will consider the evidence for this in Chapter 6). But it is worrying that we don't know at present about the validity of the rankings. Neither do we know with any certainty how the discourses of hunters' love and hate, awe and killing (achievement and appreciation) fit together pragmatically, emotionally and psychologically.

Clearly, the analysis of on-line hunting forums is an interesting approach but one that is currently fraught with difficulties when we attempt to assess the generality of the findings and their implications for wildlife management, which Ebeling-Schuld and Darimont's 2017 article does. The problem with pandering to the achievement motivation of those who boast online about their achievements in order to feed some narcissistic drive (as we shall see later) is that this can be an upward and never-ending spiral for both narcissism and narcissists, as Twenge and Campbell note in their worrying book titled *The Narcissism Epidemic* (2009). The authors do not discuss hunting or trophy hunting. They do, however, cover many other aspects of life including the use of social media to generate admiration and affirmation in our developed (and growing) 'age of entitlement'.

Of course, the results of Ebeling-Schuld and Darimont study (2017) fit in with the evolutionary hypothesis. There are a lot of *prima facie* arguments for their findings since achievement and display clearly connect to social status. But language is at times a conscious and controlled mode of communication. We might ask what the analysis of human nonverbal communication, which at times can be automatic and unconscious (Beattie and Ellis 2017), tells us, particularly in the displays of dead animals (the results of all that 'achievement') where the immediacy of the communication is paramount.

Nonverbal behaviours of hunters

Some researchers have tried to move beyond the verbal descriptions to identify the psychological satisfactions associated with hunting by analysing instead the nonverbal communications of the hunters when they pose with their prey. Of course, as has already been discussed, the harvesting of large prey is not only found in contemporary society. Some evolutionary anthropologists suggest that the displays of dead animals served as important elements in competitive displays throughout the evolution of hunter–gatherer culture (Hawkes and Bliege-Bird 2002). If displays of hunted trophies have such a formidable evolutionary history (with implications for mate selection and reproduction), the psychological motivations to engage in this form of behaviour may be more hardwired into the brain and thus less conscious, and more indicative essentially of underlying, unconscious and implicit cognitions. From this perspective, our verbally reported motivations may actually be incidental or of little consequence, although, of course, hunters may draw upon this discourse to explain their own behaviours ('Men were born to hunt big prey and show what they're capable of'). After all, we may not be *consciously* aware that our choice of an Armani suit is intended specifically to signal our financial resources to potential mates; we might just think that we like its classy look and feel.

So to this end, Child and Darimont (2015) analysed the types of smiles displayed by hunters posing with their harvested prey. Specifically, they were interested in distinguishing two of the main types of smiles. The Duchenne smile (named after the 19th century French neurologist Guillaume Benjamin-Amand Duchenne de Boulogne) is a genuine smile involving spontaneous eye and mouth muscle activation. The non-Duchenne smile is a more conscious deliberative smile that lacks eye muscle activation (Ekman et al. 1990).

Spontaneous Duchenne smiles have different onset and offset times in comparison to non-Duchenne smiles. The consciously controlled non-Duchenne smiles appear and fade very quickly (although, of course, onset and offset times are not discernible in isolated trophy images). The researchers argue that true smiles 'provide an honest, involuntary indication of pleasure' (Ekman et al. 1990; Ekman and Friesen 1982; Johnson et al. 2010) and the authors therefore used this specific type of smile to infer satisfaction under various contexts of achievement.

They analysed 5972 images of adult male hunters who posed alone without prey or with prey of various sizes (small or large) and different types (ungulates or carnivores). These photographs were taken in British Columbia and Alberta in Canada and were available on professional guide outfitter websites or online hunting forums. In order to identify the types of smiles, the authors scored AU6 (mouth) and AU12 (eye facial muscles) as either active or inactive. They focussed on Duchenne smiles in which both eye and mouth muscles were activated, checked the reliability of scoring of the types of smiles, and excluded smiles for which their confidence in scoring was low. They started with a large set of 3181 smiles; 2791 images remained in the final sample.

Their results revealed that Duchenne smiles were significantly more likely when hunters were photographed with prey than without prey and also significantly more likely when the hunters posed with large prey compared with small prey. Older hunters showed more Duchenne smiles than younger hunters. When posing with carnivores rather than herbivores, older hunters were significantly more likely to show Duchenne smiles. The authors concluded that their study produced 'independent evidence that displaying prey evoked satisfaction in some achievement context. Moreover, that old hunters actually show more satisfaction displaying large/dangerous prey than when posing with small/herbivorous prey suggests achievement-oriented satisfaction has not decreased with age' (Child and Darimont 2015: 9). They also suggested that 'Despite any potential biases in their sample set that photographs might provide a more accurate signal of satisfaction than data provided by hunters themselves, multiple satisfactions research is based almost exclusively on self-report survey and interview data....In contrast, emotions in general and involuntary "true" smiles in particular generally do not lie and offer an alternative to these traditional satisfaction ratings' (p. 10).

Child and Darimont assume here that Duchenne smiles, especially in photos showing larger carnivores, indicate more genuine positive emotions with this category of kill and therefore provide more accurate indicators of underlying psychological state than self-reports that can be consciously generated. This is an interesting and innovative study, but again the theoretical assumptions that guide the analysis and interpretation of the results are worth examining.

The first assumption is that these photographs taken from various online sites may provide more accurate signals of satisfaction than other sources. This is problematic for a number of reasons. Photographs are not unbiased snapshots of behaviour. Rather they are carefully posed images recorded for the appreciation of an intended audience. The selection aspect alone could explain any differences in distributions of smiles with respect to different species. Hunters want to look genuinely happy with large carnivores (after all, they've gone to all that trouble and expense).

Most people can generate genuine-looking Duchenne smiles for the camera. (Many people have conscious strategies for this, for example, thinking of something very positive. I just say 'monkey' and it usually does the trick.) Recording onset and offset times could have been very helpful in adding additional detection criteria to the study and perhaps even *essential* for reliable and valid identification of genuine Duchenne smiles.

Hunters want to communicate certain images to their intended audiences (mainly other hunting aficionados) about the *lived* experience of hunting. As Halla Beloff (1985: 16) says in her book titled *Camera Culture,* 'From the beginning, photography has shown the world. That is a truism. Words can describe the world. The camera has the power to authenticate'. Hunters authenticate their experiences in images of the culminations of their hunts. *Natural* and *spontaneous* smiles (carefully selected if necessary) are required for the process of authenticating the joy of hunting and killing—the natural end point of the hunt—and communicating something about themselves in costly signalling terms. They want to signal that they are relaxed, confident, and spontaneous in expressing emotions with nothing to hide while avoiding non-Duchenne (masking) smiles designed to cover negative emotions.

We must never forget that notwithstanding the point about the selection of the final image to post, having a photograph taken under any condition is a complex (if sometimes relatively quick) multi-level experience and process. As Roland Barthes states in *Camera Lucida: Reflections on Photography* (1981):

> Now, once I feel myself observed by the lens everything changes; I constitute myself in the process of 'posing'. I instantaneously make another body for myself, I transform myself in advance into an image. This transformation is an active one; I feel that the photograph creates my body and mystifies it, according to caprice....In front of the lens, I am at the same time, the one I think I am, the one I want others to think I am, the one the photographer thinks I am, and the one he makes use of to exhibit his art.

The camera can capture *natural* behaviour where smiles are much less frequent, and this is seen in the work of certain photographers like Richard Avedon. Posed photographs, however, are more self-conscious and deliberate communications and not necessarily readily useable as evidence of nonverbal leakage of real emotions (Ekman 1985; Beattie 2016).

The communicative power of the visual image

The Duchenne smile of a hunter with large prey such as a lion or rhino at his feet conveys a powerful, immediate and readily interpretable cultural signal about authority, dominance, financial resources and entitlement. In analysing how men and women are portrayed together in another form of visual image, commercial ads from the late 1970s, sociologist Erving Goffman in his book *Gender Advertisements* discussed how nonverbal communication becomes hyper-ritualised in these ads. The advertisements are meant to be seen as *natural* representations, but Goffman noted that advertisers conventionalise our conventions. In the 1970's advertisements shown in his book, men are shown as larger and more dominant, doing important jobs (*function ranking* according to Goffman), standing rather than sitting, eyes focussed forward. Women, on the other hand, are portrayed in more playful poses, less serious, child-like, they touch their faces more.

According to Ekman and Friesen (1969), face touching is a form of self-adaptor used for self-comforting; both men and women use face touching in everyday life but only women tend to use them in advertisements. Women in advertisements often appear to be looking away. Goffman called this 'licensed withdrawal' to indicate a woman is happy to be guided by a forward-looking male figure. He concluded that these advertisements display the rituals of subordination. Halla Beloff (1985: 238) wrote 'The interpretation of the demography of photography is that women are seen essentially as children. They have the rights and privileges of children, and lay the same price as children for that position....If these are her privileges, she is therefore subjected to control'.

Of course, many aspects of the portrayals of men and women have changed since the late 1970s, but the substantive point remains: many genres of visual images need quick interpretation, without much conscious reflection (Kahneman 2011), and this is assisted by hyper-ritualising aspects of everyday nonverbal communication, exaggerating and stylising the behaviour. This is not a trivial process; it is used to propagate psychological and political messages about, power, position and privilege.

The natural-looking Duchenne smiles that appear in images of trophy hunters should not necessarily be viewed as natural nonverbal signs of achievement. They are hyper-ritualised displays ('Just think of what you've managed to bag here and what the folks back home are going to think about this. Wait till they see this!'). Perhaps, the most interesting aspect of the frequency of Duchenne smiles in the Child and Darimont study (2015) is the converse—the relative infrequency of non-Duchenne smiles, the other side of the coin so to speak.

Trophy hunters have no need to mask smiles to cover up negative emotions. Their photographs reveal no moral or ethical qualms or regrets to be hidden by faint mouth-only smiles as they sit hunched over dead animals. The mouths of dead lions are sometimes propped open to show their teeth to demonstrate how powerful and dangerous they are. Hunters often hold dead ibexes by their horns to keep the animals upright or hold other dead animals by their ears. These images send psychological and political messages about power and dominion over all living things and about the joy of exercising that power and doing what comes *naturally* to men. Isn't that *natural* activity the result of evolution? (they would say).

What about women? What role do they play in these *natural* celebrations of evolution? They appear to have taken the Goffman concepts of function ranking and subordination and subverted them. Female hunters often appear more dominant than men when photographed with their animal trophies. Kendall Jones was photographed astride the lion's back, pulling its head back by the mane. Another photo shows her standing with her boot on the lion's rear, hand on hip, crossbow in hand, and beaming Duchenne smile. The evolutionary justification only gets you so far.

The images propagate the basic psychological, political and indeed cultural messages of trophy hunting—power and privilege and entitlement without moral uncertainties. These great species become props in stories about hunters, their great achievements, values, rugged individualism, pursuits of freedom, and their resources. These images are supposed to be admired. But how can anyone possibly justify trophy hunting activities? That's where language comes in, as we shall see.

SUMMARY

1. In everyday life, deep and unconscious influences on human behaviour may not be easily or readily articulated by the actors themselves. We considered two domains in which this is clearly the case: sustainability and racial bias. We must be cautious when accepting reported motivations for any actions, including trophy hunting, without further consideration and analysis.
2. People give meaning to their actions through language, so I reviewed studies that analysed posts on hunting websites. The main *satisfactions* that hunters derive from this pastime are, according to this research, *achievement* (feelings of satisfaction related to performance) first and foremost, followed

by *appreciation* (enjoyment of the experience) and *affiliation* (strengthening of personal relationships or enjoyment of the company of others).

3. I suggested that these analyses are lacking in certain areas by excessive focus on individual phrases and words, and suggest an alternative discourse-based (top-down) functional perspective to better analyse hunters' accounts.
4. Some researchers have tried to move beyond verbal descriptions to identify the psychological satisfactions associated with hunting by analysing the nonverbal communications of the hunters when they pose with their trophy animals. Some evolutionary anthropologists suggested that the displays of dead animals served as important elements in competitive displays throughout the evolution of hunter–gatherer cultures. I suggest that the psychological motivations to engage in this behaviour may be much less conscious and that the nonverbal communications of trophy hunters may well be more revealing. From this perspective, verbally reported motivations may actually be incidental or of little consequence, although, of course, hunters may be able to draw upon this discourse to explain their behaviours ('Men were born to hunt big prey and show what they're capable of').
5. The research on the nonverbal communications of trophy hunters revealed that *spontaneous* and *genuine* Duchenne smiles are far more common when hunters are photographed with prey than without prey and also significantly more likely when the hunters pose with large prey compared with small prey.
6. Older hunters appear to show more Duchenne smiles than younger hunters.
7. When posing with carnivores rather than herbivores, older hunters are significantly more likely to show Duchenne smiles.
8. One conclusion of the authors cited in this chapter was that their research produced 'independent evidence that displaying prey evoked satisfaction in some achievement context. Moreover, that old hunters actually show more satisfaction displaying large/dangerous prey than when posing with small/herbivorous prey suggests achievement-oriented satisfaction has not decreased with age.'
9. I critiqued this work by arguing that these Duchenne smiles are not necessarily always natural nonverbal signs of *achievement* and may represent hyper-ritualised displays of specially selected expressions to be posted on social media. The mouths of dead (and formerly ferocious) animals are sometimes propped open to show their teeth. These images send out psychological and political

messages about power and dominion over all living things and about the joy of exercising that power, and doing what comes naturally to men.

10. The images propagate the basic political and cultural message of trophy hunting: power, privilege and entitlement without moral uncertainties. These great animal species serve as props in stories about hunters and their great achievements, values, rugged individualism, pursuit of freedom and, of course, their resources.

Chapter 4

Justifying the unjustifiable?

The rhinoceros carefully arranged as a prop.

I ask people why they have deer heads on their walls. They always say because it's such a beautiful animal. There you go. I think my mother is attractive, but I have photographs of her.

Ellen DeGeneres

Ellen DeGeneres is a very well known American comedienne, television host, actress, writer and producer. She has hosted a syndicated television talk show since 2003. She is noted for her wit and pithy one liners. This chapter is about how hunters *justify the unjustifiable.* The quote from DeGeneres provides us with a stark frame as we start to consider this issue.

For many people faced with images of dead lions or dead rhinos and smiling hunters (regardless of the type of smile or its psychological origin), what immediately comes to mind is how the hunters could *possibly* justify this killing and wanton destruction of life. It is interesting that when asked to justify trophy hunting, Kendall Jones and other big game hunters often suggest that justification is unnecessary ('I have a hard time with the word "justify" because it implies that I am doing something wrong and have to make excuses for my action').

Social scientists have analysed how people justify and excuse various types of actions that are open to strong criticism (and moral outrage) and identified a taxonomy or descriptive classification of justifications and excuses used in a variety of contexts when discussing many types of blameworthy activities. It may be useful to consider the similarities and differences that appear in the justificatory techniques used by various groups, to see how justifications and excuses differ among groups and to determine what, if anything, they have in common.

Justifications and excuses

Social scientists argue that justifications and excuses are different types of 'accounts' (although Kendall Jones thinks that they are one and the same as she uses the terms interchangeably). Scott and Lyman (1968) defined accounts as 'statements made to explain untoward behaviour and bridge the gap between actions and expectations'. Based on Austin's (1961) formulation, they suggest that accounts may be classified as either justifications *or* excuses, depending upon their specific content and reasoning.

An excuse represents a denial of personal causation—intentional or otherwise. On the other hand, to justify an act is to recognise that it is somehow untoward but necessary given the context within which it occurred. Thus, a person who justifies an act admits personal causation, but in so doing provides a reason for behaving as he or she did. Scott and Lyman (1968) identified a number of subtypes of justifications and excuses. They outlined four main justificatory techniques:

1. Denial of injury: The actor denies that anyone (significant) was seriously or otherwise injured by his behaviour. A denial of injury can still be maintained in the face of evidence to the contrary if the victim is constructed in the account as insignificant or deserving of the injury.
2. Denial of the victim: The action is justified by arguing that the victim deserved the injury (e.g., if the victim previously assaulted the actor).
3. Condemnation of the condemners: The actor recognises the impropriety of his or her behaviour, but argues that it is insignificant in comparison with how others behave and what they get away with.
4. Appeal to loyalties: The actor maintains that his or her behaviour is acceptable as it is in the interests of another individual or an organisation that has authority over the actor or to whom the actor owes some type of allegiance.

Excuses, on the other hand, are different, but they too can be subdivided into four main categories:

1. Appeal to accidents: Responsibility for an action may be denied by citing some generally recognised hazard or set of exceptional circumstances as the cause. This can also apply to an understandable loss of control over the actor's motor responses.
2. Appeals to defeasibility: An actor may claim that he or she did not act under *free will*, acted without full knowledge of the circumstances surrounding the behaviour, or was unaware of the consequences of the behaviour.
3. Appeal to biological drives: The actor may excuse his or her behaviour as the result of some inherent biological capacity over which he or she does not have full control. An example may be a politician who, after punching a protester who threw an egg at him, claimed that the assault was a reflex action committed without thought. (A famous historical example involved former UK Deputy Prime Minister, John Prescott.)
4. Scapegoating: The actor may claim that he or she behaved simply in direct response to the perceived attitudes or behaviour of another.

These classifications serve as useful tools for examining the accounts of trophy hunters. Given the intentionality of trophy hunting, it is clearly difficult to use excuses. Trophy hunting is not accidental; it is a free will choice. Trophy hunters clearly have control over their biological drives to kill. Where would uncontrollable biological drives leave society as a whole? Trophy hunting is clearly not a direct response to the specific behaviours of the animals in question since the hunters travel to specific locations at great expense with the intention of killing animals.

Trophy hunting is a behaviour the participants must justify (no matter how inconvenient that is for Kendall Jones or anyone else). Denial of injury is hardly possible based on photographic evidence of hunters posing with dead animals. However, the other three types of justifications can and do operate to varying degrees. Hunters can (and often do) argue that the animal deserved to die ('It was a rogue animal, aggressive, terrorising the neighbourhood'). This, of course, is a form of victim blame. Hunters also condemn the condemners ('They don't manage the animals properly; a cull is necessary; I've had to travel all this way to help you guys out'). Hunters also appeal to the loyalties of like-minded individuals who share their values of rugged individualism and independent thought.

It could be very informative in an effort to contextualise the justification techniques of trophy hunters to see how other people who routinely engage in behaviour involving violence and suffering also justify their actions. In particular, it may be helpful to consider the justificatory practices of those who engage in types of one-sided violence in a variety of unequal contests (and trophy hunting is certainly an unequal contest). Examples are highly trained nightclub bouncers who encounter often inebriated customers and heavily armed terrorists who target unarmed civilians. We might examine their behaviours to learn about the principles and practices of justification they utilise.

A number of years ago, a postgraduate student Robin Jordan and I interviewed a group of nightclub bouncers who all had extensive experience working as doormen at nightclubs and pubs in the North of England (Jordan and Beattie 2003). These bouncers were professional 'hard men'; a number had been professional boxers or trained to very high standards in martial arts. They routinely engaged in violence as part of their jobs. How can they possibly justify their violent behaviours when confronting ordinary punters?

Justifying unequal interpersonal violence

We interviewed the bouncers at various locations, and were keen to put them at ease so that they would talk freely and openly. We asked them to talk about any aspects of the violence they witnessed whilst working on doors. When we started to analyse their accounts, our first observation was the structure of the accounts. The doormen drew upon various 'grammars' of action—prescriptive rules of violence (see Auburn et al. 1995)—when constructing their accounts: (1) the violence (even when extreme) they perpetrated was necessary in order to successfully undertake the role of a doorman; and (2) violence perpetrated by others ('troublemakers') was unnecessary and irrational (disorganised) and therefore unjustified and inexcusable.

Other descriptions in the accounts seemed to serve different but complementary functions. For example, the violence perpetrated by the doormen appeared to be constructed in a way that downgraded its severity and intensity and the resulting injuries sustained by victims (if such injuries were mentioned). Most of the accounts culminated in success for the doormen involved; in other words the doormen came out on top and were often portrayed heroically. Although some of them sustained 'one or two minor injuries in the fracas', they nonetheless were successful in, for example, stopping a group of 'troublemakers' from gaining entry to their clubs. In fact, of all the accounts of violence the doormen reported, fewer than 16% ended with the doormen being overpowered.

Considering the relative frequency of violence experienced by these doormen, it seemed surprising that they never sustained more serious injuries. One possible explanation is that a doorman who admits that he regularly comes off second best in encounters, thus positioning himself as an occasional victim, detracts from his role as someone paid to keep the peace and is capable of doing so when necessary. We were also not simply told that 'X did this, so I hit him'. We often received highly detailed scene-setting information preceding the actual descriptions of the violence.

We noticed one interesting recurrent feature of these accounts early on: the way the doormen bolstered the *objectivity* of their accounts by drawing upon information about the past behaviours of the victims (*consistency*) and describing how other doormen would have reacted in the same way (*consensus*). Their accounts clearly indicated that any violence on their parts was *distinctive* to a certain individual or set of individuals.

This links well with a core theory in psychology. In 1967, Harold Kelley, introduced a co-variation model of how causality for events is assigned in everyday life in an area of psychology called attribution theory. He used three basic dimensions ('consistency', 'consensus' and 'distinctiveness') in a sort of calculus of blame assignment. Below is an example from a bouncer called Geoff. This extract, transcribed as said (a guide to the transcription appears below), is drawn from Geoff's account of an incident in which his refusal to admit a group of men to the club led to violence involving shots being fired and CS gas being sprayed.

> What happened then (0.2) was a group of men (2) °from a certain area came° (2) an- ↑these guys had been ↑ba:rred, they must have been knocked back from everywhere else so they've come to the pub (0.1) where (0.2) I was working (.) and a friend of mine (0.2) was working °there°. AND I KNEW one or two of them you know like (0.1) °so called bad boys° (0.3) whatever they want to call themselves. [Note: numbers in brackets indicate pauses in tenths of a second.]

The transcription below was taken from Potter (1996: 233–234) and adapted from Jefferson (1985):

1. C: Was that the time that you left? =
2. W: =He left the:n that was — [nearl] y two years ago.
3. C: [°Yeh°]
4. W: He walked out then. Just (.) literally walked out.
5. (0.8)
6. C: ↑Oka↓y. So, (0.5) for me list↓enin:g, (.) you've
7. got (0.5) rich an:d, (.) complicated lives,
8. I nee:d to get some his [tory to put —]
9. W: [Yyeh. mmm, =]
10. [Mmm. (.) Ye:h (.) Oh ye:h.]
11. H: [=Yeh. (.) that's (.) exactly wha]t ih °um°

The symbols used in the above transcript are detailed as follows:

- Underlining (walked out) indicates words or parts of words stressed by the speaker.
- Colons mark prolongation of preceding sounds (the:n); more colons indicate longer prolongations (Ah:::).
- Arrows precede marked rises and falls in intonation (↑Oka↓y).
- The question mark in line 1 indicates questioning intonation (there is no necessary correspondence with utterances participants treat as questions).
- The full stop (for example, in line 2) marks a completing intonation (not necessarily a grammatical full stop).
- The comma in line 6 marks a continuing intonation (not necessarily a grammatical comma).
- A hyphen (for example, Thanks- Tha:nksgiving) marks an abrupt and noticeable termination of a word or sound.
- The brackets in lines 2 and 3, 8 and 9, 10 and 11 mark the onset and completion of overlapping talk.
- Statements that run into one another are marked by equal symbols (lines 1 and 2, 9 and 11).
- Numbers in brackets (0.5) indicate lengths of pauses in tenths of a second; a full stop in a bracket (.) indicates a pause which is hearable but too short to measure.
- Talk that is quieter than the surrounding talk is enclosed by degree symbols: °yeh°.
- Talk that is louder than the surrounding talk is fully upper cased (WHERE).
- Clarificatory comments appear in double brackets: ((laughs)).

Geoff states that the group of men in question 'must have been knocked back from everywhere else'. That suggests that doormen at other pubs and clubs in the area had problems with this particular group and consequently the group was barred from other establishments.

In Kelley's terms, Geoff's statement invokes the consensus dimension by aligning his experiences with the group and those of other doormen in the area. Since the group of men had been barred from other establishments—not just the one where Geoff worked—we can make the assumption that these men were known troublemakers and recognised as such in the doorman community. Distinctiveness, while not explicitly mentioned in this section of the account, can be inferred to be at high level since it is implicit throughout Geoff's narrative that he does not engage in violent confrontations with all groups of young men come to the bar where he works.

Indeed, referring to this same incident, Geoff goes on to say that, 'The best thing to do in these situations is just to leave them, or try and keep people away from them, and that's it basically.' Finally, consistency can also be said to be high since we are told that these men had previously been barred from Geoff's establishment, although we are not told why. Thus, Geoff and his colleagues found the men's behaviour unacceptable on prior occasions.

We can see how the specific combination of Kelley's variables—high consensus, high distinctiveness and high consistency—is made available to us 'objectively' within the account so that we as listeners can easily make a certain type of attribution—blaming the group of men for the onset of the violence rather than attributing the violence to the doormen (for example, their potentially provocative behaviour with Geoff and his colleagues to refusing them entry to the bar) or the situation (the club, the alcohol, the time of night and other factors). This is exactly what Kelley's covariation theory predicts.

In terms of the language used, Geoff describes the group who were previously barred, returned on this occasion and attempted to gain access as 'a group of men from a certain area'. This description works effectively at a number of levels. It builds the possible threat level posed by this group. Mention of 'a group of men' hints at an organised unit in a way that another description ('a bunch of lads') might not. It is also significant that we are not told precisely where these men are from; we are offered the vaguer description that they came 'from a certain area'. This makes the men appear sinister, and possibly from an area notorious for violence.

At this point it may be useful to consider what the detailed transcription of the language used actually brings to our analysis. In the above extract, for example, before Geoff informs us where the group of men are from, a significant pause (2 seconds) indicates that he is considering whether to divulge this information, and if so, how. This is an interesting feature of this section of Geoff's account as the pause operates similarly to a strategy identified by Scott and Lyman (1968) to avoid giving an account or in Geoff's case, not elaborating on certain details.

Mystification (Scott and Lyman 1968) can be employed as a way of offering a limited amount of information and suggesting that although there may be more of significance to say, this is not possible for one reason or another. In this particular context, it may be that Geoff wishes to avoid being held responsible for providing enough detail that the men could be identified. The avoidance is understandable given that the incident involved use of a firearm. Confirmation comes when we are subsequently told only that 'a group of men from a certain area came'. Note also, as indicated in the transcription, this is said in rather hushed tones which supports the argument made earlier that the inclusion of vagueness has the effect of aiding in Geoff's construction of these men as threatening and dangerous to the extent that revealing where they came from could lead to their identification. His vague statement reveals their notoriety and suggests possible retribution.

Since the incident described by the doorman ends with firing of gun shots, use of CS gas, and subsequent imprisonment of the doorman on charges relating to the incident, this analysis may not be far-fetched. It is also interesting that Geoff does not simply describe the men as 'bad boys'. He attributes that description to the group of men. This aids in the overall construction of these men as more than persistent troublemakers, but as troublemakers who revel in their reputation for violence, making their behaviour on this occasion seem not unexpected.

The fact that Geoff lowers his voice as he describes the men as 'so-called bad boys' might also indicate that he is not wholly in agreement with that description. Put another way, even if the group regarded themselves as 'bad boys', Geoff does not seem to agree. The inference is that as far as he is concerned, they are nothing more than an irritation. Therefore, the construction of the social context of violence in this way facilitates the negotiation of blame—even for proactive acts of violence—away from the doorman and onto the others by the rhetorical construction that depicts them as blameworthy victims.

The second extract concerns an incident when a doorman (Mick) was called to a bar because a man inside was behaving threateningly towards the bar staff and other customers:

> He was an ex-para. () a big kid () who comes round all the time looking for trouble. (0.5) he does—he does it all the time. he's known (0.3) when he's out↑ with his wife and that (0.2) he's as nice as pie, (0.5) but as soon as he gets out (0.1) he-he doesn't go back home—he stops out drinking. () after two or three pints (0.2) he's SMACKing his mates. He's a nut.

A key feature of this account is Mick's assertion that the man in question 'comes round all the time looking for trouble. He does—he does it all the time.' What is significant about this isn't simply its obvious characterisation of the man as a known troublemaker, but rather its vague formulation of exactly where the man goes. When Mick tells us that the man 'comes round all the time looking for trouble', does he mean that the man goes to the club where Mick works, or does Mick mean that the man goes to that general locality and hence a number of clubs and bars in the area where he developed a reputation as a troublemaker?

Mick is effectively saying that both *consensus* and *consistency* within the account are high, as both Mick and other doormen in the area had trouble with this man on other occasions. As in the previous example, we may reasonably infer that distinctiveness operates implicitly within the account since Mick does not hit anyone who comes in contact with him at the bars and clubs where he works. Indeed, by definition, Mick's account covers an exceptional or *distinctive* event. Thus we have the necessary combination of Kelley's variables that allows us to attribute blame for the violence to the other person (something about him) rather than the doorman on the basis of the *background* and again done in an *objective* manner.

As noted earlier, specific language can be quite informative. The first piece of information Mick provides about the troublemaker is that he was an ex-paratrooper. Mick could have drawn on many biographical descriptors (married man, father of young children, dog lover, mechanic) but didn't. Mick refers to the troublemaker's previous occupation (paratrooper) right at the beginning. This descriptor is introduced to manage and direct our image of the type of person Mick faced: tough, resolute, presumably physically strong, capable of looking after himself in a fight (without explicitly spelling out the associated features). Such characterisation has the effect of working up the threat posed by this man to Mick and the other people in the bar, in turn making Mick's subsequent use of violence seem reasonable and even necessary.

It is also worth noting that Mick is specific about the regiment to which this man belonged. Rather than simply telling us that the man was an ex-soldier, he notes that the man was a member of one of the elite units in the British army. The inclusion of such detail serves not just to bolster Mick's overall construction of this man as potentially threatening, but also to underline the nature of the threat as being more serious than that posed by an average troublemaker. Furthermore, Mick's construction serves a dual function. It paints a picture of this man as having the potential to pose a significant threat to both Mick and other customers, in turn casting him as a deserving victim. At the same time, it operates at a subtler level: despite this man's reputation, Mick can handle the situation should it become violent, as it did later in Mick's account. Furthermore, the account implicitly flags up his professional capabilities.

The next piece of information that Mick provides is the man's size. Mick describes the man as a 'big kid'. This is interesting because on the one hand it helps establish the man as a significant threat ('big') whilst describing him as a 'kid'. The phrase may be a colloquialism but nonetheless hints at a certain power imbalance. Mick would not describe himself as a 'kid' in such accounts. In other words, even though the man is a former paratrooper with a history of threatening and violent behaviour, he is still no match for Mick. The constructions of *self* and *other* go hand in hand in these sorts of accounts.

Finally, Mick describes the man as 'a nut'. Mick tells us that this man's behaviour tends to become more violent with the consumption of alcohol and the escalation can happen after a relatively small intake ('after two or three pints he's smacking his mates'). Therefore, the man's behaviour is not simply a result of alcohol use, but rather is attributable to his intrinsic characteristics. There's something about *him*, some flaw in the man's character that is responsible for his troublesome behaviour beyond the effects of small amounts of alcohol. Mick stresses the *smack* in smacking as if to underline the unreasonableness of this behaviour. The stress serves to establish Mick as a person capable of recognising and differentiating between someone who is aggressive because he is drunk, from someone who is aggressive because of his personality.

The violence deployed by a doorman to deal with a drunken troublemaker whose behaviour may be looked upon as more excusable might be a lot less severe than that meted out to an aggressive person. Thus, Mick implicitly provided us with information as to his own character: he is a reasonable type of person who does not necessarily engage in violence simply for the sake of it.

One other feature of the scene that Mick set before describing the ensuing violence also relates to the establishment of the troublemaker as an unreasonable and unbalanced person. Mick notes that the ex-paratrooper can be 'nice as pie' when he is out with his wife but is also capable of hitting his own friends after a few pints. This constructs the other person as unpredictable and volatile whilst it constructs Mick as a fair and balanced observer. He is prepared to identify the positive aspects of another person's character by citing good points. In other words, he is showing us that he is not biased and presents a fair account of what he witnessed. He tells us that the other man in the altercation was 'as nice as pie' when out with his wife. At the same time, Mick conjoined the 'nice' description to statements about the other side of the man's character.

There are a number of general points to be made at this point about the processes and uses of justification. The first is that when we think of justifications and justificatory statements in everyday life, we often think of short self-contained responses. The taxonomic approach to justifications proposed by Scott and Lyman (1968) does not dissuade us from this thinking; it may even encourage us to think about justifications in this exact way. Anyone listening to highly trained professional doormen talk about the violence they inflicted on less trained, unequal and often inebriated punters will be struck by the way the justification is built into the description of the episode, even into how the episode is introduced.

Objectivity in a justificatory account is bolstered by using a calculus of reasoning (identified by Kelley in attribution theory terms) about consistency, consensus and distinctiveness to point our search for causality in one direction or another. After all, the violence involving doormen could be attributed to the doorman or the punter, to the situation, or to some complex interaction of people or events. The accounts of the doormen employ this calculus to put the blame on the dispositional characteristics of the other person (troublemaker) rather than on themselves, alcohol use, the specific situation or any other factor.

Scott and Lyman would say that the justifications seen here are based principally on denial of the victims. These are worthy victims—volatile, disturbed troublemakers, capable of real violence who revel in their reputations, victims who deserve all they get. Conversely, we find a very different construction of self by the doorman: the troublemaker is a professional (former paratrooper) and I'm a professional. I can admit that the man can be as nice as pie. I'm fair-minded. I only smack troublemakers or sinister so-called bad boys from bad areas.

Despite all of this blaming of victims, at no stage in any of the doormen's accounts do they detail injuries sustained by victims. Indeed, in one account, a doorman named Gary described how a fight erupted between a group of professional ice hockey players and the doormen of the club which left at least one of the ice hockey players unconscious. When asked whether anyone had been hurt, Gary answered that, 'Mick [a fellow doorman] broke his finger and that were it' (sic). No mention was made of the injuries that were presumably sustained by the individuals who fought with Gary, Mick and their colleagues. While the doormen often said that their opponents were rendered unconscious following a punch or punches, they never mentioned associated injuries.

In another extract, Mick tells us that 'Anyway I bomped him and he went down. I kicked him all over and the next day I couldn't walk. I thought I'd broke my feet' (sic). Note that despite having punched and kicked the ex-paratrooper unconscious, Mick positions himself as a victim of sorts in that he describes how badly his feet were injured as a result of his actions. Once again, no mention is made of the injuries sustained by the other man.

This failure of the doormen to acknowledge such injuries beyond alluding to the victims' states of unconsciousness—sliding down walls, feet laid out under a table—in an almost comical way serves to downgrade the violence and its consequences while at the same time highlighting the efficiency of the doormen in dealing with troublemakers. The descriptive accounts of these one-sided violent incidents by professional hard men contain interwoven justificatory strategies.

The transformational power of language in the hands of terrorists

I have also interviewed terrorists who utilised similar justificatory moves when they talked about what they did during the Troubles in Northern Ireland. The Troubles that spanned three decades concerned the political future of Northern Ireland. On one side, the Republican movement aspired to a United Ireland. The Unionists on the other side wished to maintain the union with the rest of the United Kingdom. Political aspiration largely corresponded to religious affiliations: Protestants (mainly Unionists) versus Catholics (mainly Republicans), with various paramilitary organisations waging a guerrilla war against their opponents. Of the 3500 people killed, 52% were civilians and 16% were members of various paramilitary groups. The paramilitary IRA and INLA on the Republican side and the UDA and UVF on the Unionist side claimed that they were not targeting civilians in the conflict and that they were engaged in a war against their enemies.

The example below involves the murder of a random and entirely innocent Catholic by a loyalist paramilitary member known as Hacksaw and his gang. The victim was driving along Cambrai Street off the Shankill Road in North Belfast with his 10-year-old son beside him. I listened as Hacksaw explained the 'targeting' of known enemy combatants and the 'intelligence' his group used to guide targeting and murder. I asked Hacksaw whether the murdered man was someone his group *definitely* had intelligence on rather than a random Catholic. Hacksaw said:

> That was a random because we missed the one we were looking for. So this guy came along and we just wiped him out. I know it's sad. But that night I felt like he had to go, he just had to go. You see, we'd taken the chances of hijacking the car, going way up the Shankill with the guns in the car. We weren't going home empty handed. Because if we had been caught we were getting big time, so he had to go, instead of us going home empty handed.

Hacksaw turned himself into the victim ('we were getting big time') and suggests he had no choice (the victim 'had to go'). The victim was treated as a thing ('a random'), depersonalised, without a name, a substitute ('we missed the one we were looking for') in what Hacksaw saw as a great act of bravery.

Like the hard men who tend bar and club doors, terrorists don't mention the damages they inflict on their unequal victims. They minimise, neutralise and downplay such injuries. Here's Hacksaw again:

> So this guy came along and we just wiped him out. [How do you 'wipe out' a human life, as if it were a stain?] We let him get to the ramps on Cambrai Street and moved out in front of him and shot him. He was

> in a car, his 10-year-old son was with him, but I can put my hand on my heart and say that I didn't see the child. The child was sitting on the passenger side down, but apparently the child was hit in the leg but we didn't see the child. The next day when I heard about this I wasn't worried about the father, but I was interested in the child, but when they gave it out that the child was only grazed I said, "Well, that's happy days, the child is alright." But I wasn't worried about the dad because that was one less to worry about.

The child's physical and psychological injuries were glossed over, downplayed, not discussed. Although the child witnessed the murder of his father and was also shot, Hacksaw reframed the situation as 'happy days, the child is alright'. Self-construction is a critical component of justification: 'We didn't see the child' can be construed as 'Do you think that I'm the kind of person who shoots children?

How do you *justify* acts of random terrorism against civilians? By building your whole account around it; you don't ring fence it; you construct yourself as the victim ('if we'd been caught...'), and denigrate the real victim. [Hacksaw claimed 'they' denied the victim was in the IRA, but implicitly 'they' deny a lot.] Justification means never dwelling on the injuries inflicted. Terrorists believe their murders are 'clean'. Hacksaw's group 'wiped out' a random victim. The doormen described their victims as 'sliding down the walls', 'feet laid out under a table' in attempts at humour. People engaged in unequal contests try to justify their actions by feeling sorry for themselves, Mick, the hard man at the club door, suffered sore feet after all that kicking of his victims. Hacksaw felt nervous about being stopped by the police or army. Mick hurt his finger knocking one of his victims out.

The nature of justifications in the language of trophy hunters

It seems that justifications are ubiquitous in certain types of accounts. They are highly structured. We don't really need to ask Kendall Jones for a justification; she provides it throughout her interviews about trophy hunting. She constructs herself as a victim:

> Why would these huge, powerful organizations [like the International Fund for Animal Welfare] go after me, a woman....I am not the first to go on African safaris yet these groups attack me nonetheless.

She evokes the dimension of consensus to again deflect blame:

> I am not the first to go on African safaris.

She denigrates the victim:

> They only permit the hunting of five rhinos a year. Normally these are rhinos that are old and no longer able to reproduce, sick, or too dangerous to coexist with the other rhinos.

She implies that such rhino victims are *bad* and don't even get on with other rhinos. She creates a consensus with the *good* rhinos to ensure that the *bad* rhinos that are hunted get 'what's coming to them'. She uses that odd calculus from attribution theory to locate the causation of the violence and place blame fairly and squarely on the victim. Then there's the act itself:

> Hunters have to actively track, chase, and engage the animal in order to have a successful hunt. It is dangerous, life-risking and difficult. [Even with old and infirm animals?]

The terrorist tells us how much he cares for children and worries about their welfare after his gang shot a child. Doorman Mick shows that he's fair and decent and will always call a victim a 'good lad" even if he kicked the good lad to the point where his feet are sore. Kendall tells us how good she is more directly than the terrorist and doorman did:

> It also takes a lot of patience and skill to operate a rifle and even more to operate a bow.

The accompanying image depicts a beaming and heavily made-up Kendall with her bow and arrow on the rear of a dead lion:

> It's a very connected and even spiritual experience and you learn a lot from yourself. It's definitely helped me grow as a person.

It would be very odd to hear the terrorist who undoubtedly learned a lot about himself in his years on the streets in Belfast finish with 'It's helped me grow as a person.' At least, *he* didn't say that.

Kendall also uses another justificatory move identified by Scott and Lyman (1968). She condemns the condemners throughout this interview and others. She truly appreciates the animals she kills and the condemners don't. She has the skills and courage to enjoy the hunt; the condemners would find the hunt 'unnerving'. She is fully informed and they are misinformed. Hunting has made her compassionate:

> That compassionate attitude is why I donate my game meat to the local villages'.

The condemners are not compassionate. They even attack her, a defenceless woman, who deserves none of their condemnation.

Justifications provide us with structures for our narratives and stories. They give meaning to our lives. They emerge freely as we talk about what we've experienced and what we've done. There are clearly common elements in the accounts of the hard men on the door, the terrorists on the streets in Northern Ireland, and the trophy hunters in Africa. The logical calculus of attribution theory is employed to objectify the assignment of causation and blame, the construction of self and victim, and the downplaying of the injuries they inflict. I have now looked at several thousand images of dead trophy animals. Very few of the pictures showed the actual wounds. It's funny that.

We are creatures of language. It's what makes us human and separates us from all other species. Language allows us to explain. Without language and justifications, we would be lost. I recalled that although Hacksaw found religion in prison, he nevertheless attempted to justify the nature of what he had done in a complex and insidious manner. He murdered a completely innocent person who simply found himself going in the wrong direction in the wrong place at the wrong time. If the victim had been driving in the opposite direction, he would have survived because Hacksaw would have presumed that the victim was a Protestant.

We need to view justifications for what they really are. We use them habitually in everyday descriptions, often with (seemingly) little conscious awareness of what we are doing. The *whole* discourse of trophy hunting needs to be considered to reveal how these pervasive justificatory techniques are employed. And, of course, justifications are not just designed to persuade other people about events that happened and why they happened. They seem to work well for the benefit of Hacksaw, the doormen and the trophy hunters themselves. Language helps us communicate with others but also helps us to give meanings to our own lives *privately*. These two processes are, of course, intimately connected as we will see in the next chapter.

SUMMARY

1. In this chapter, we explored how hunters justify their hunting by deconstructing the general processes of justification in a number of domains, all unequal in various ways, including accounts of hard men working at the doors of nightclubs, terrorists, and trophy hunters.
2. I suggest that all three accounts contain common elements including a logical calculus of attribution theory employed to objectify the assignment of causation and blame, the construction of self and the victim, and the downplaying of injuries inflicted

by the protagonists. I noted, for example, that very few trophy hunting images show the wounds of the dead animals.

3. The chapter argues that we are creatures of language. It's what makes us human and separates us from all other species. Justifications have their own unique and critical grammars.
4. I suggest that justifications may not be designed only to persuade other people about what happened and why it happened; they also seem to work for the protagonists themselves.
5. Language enables us to communicate with others; it also helps us to *privately* give meaning to our own lives.

Chapter 5

Why trophy hunters smile with such relish

A female hunter, nice open smile.

> **Truda Hangnail was smiling, and it was not the sort of smile designed to make anyone feel happy.**
>
> Vivian French
> *The Bag of Bones*

Vivian French is a popular British author with a prodigious output of novels and nonfiction for children and young adults. Like many authors before her she has commented on smiles and their meanings. Trudy Hangnail, the smiling woman cited above is a very wicked witch.

In this chapter, we consider smiles and what they mean, and it is good to be reminded that different types of smiles reflect different psychological states and have quite different functions.

In Chapter 3 we considered how Child and Darimont (2015) used the analysis of the *genuine* Duchenne smiles of hunters as a means to gain insight into their psychological motivations. The authors considered such smiles 'a more accurate signal of satisfaction than data provided by hunters themselves. Multiple satisfactions research is based almost exclusively on self-report survey and interview data....In contrast, emotions in general and involuntary "true" smiles in particular generally do not lie and offer an alternative to these traditional satisfaction ratings' (p. 10).

I already discussed their conclusion that Duchenne smiles, especially in photos of trophy hunters with large dead carnivores, reveal genuine positive emotions related to this category of kill and therefore provide better indicators of underlying psychological states. The authors reasoned that such smiles in photographs are spontaneous, more genuine, more unconscious and less open to being faked than self-reports. The counter-point I made was that the smiles depicted were *selected* and consciously encoded images (encoded for the camera) and designed to communicate information about the hunters and their relationships with the dead animals. The photographs were intended to appeal to a peer group to draw admiration and praise. In Chapter 6, we will consider why some people use killing animals as a way of satisfying this particular need and we will consider what psychological characteristics may drive them.

Perhaps it is worthwhile for us to consider smiles in more detail here for one important reason. The smiles of hunters who pose beside animals they killed often represent one of the most offensive aspects of trophy hunting for critics of the practice. How could anyone enjoy killing these beautiful animals and then exhibit their enjoyment to such a degree? When you study genuine spontaneous smiles in discourse (rather than smiles in photographs), they often reveal psychological complexities in disconcerting ways.

Returning to my interview with Hacksaw, the Irish terrorist, I asked him where he got his nickname. 'It's a long story', he said. 'It started in the Troubles when I got a bad reputation.' 'For what?' I asked. 'Bad things' was his reply and he smiled spontaneously in a way that might well have been classified as a Duchenne smile.

Hacksaw was directed to shoot dead a young person who was targeted by a paramilitary organisation, but he [kindly?] used a hacksaw to saw off some of the youth's fingers. The resulting moniker stuck. Many saw the smile of Hacksaw in that interview (filmed by the BBC and shown in its *Home Truths* documentary series) as that of a psychopath enjoying his infamy and reliving his experience with relish. But is that what the smile reflected?

That wasn't the only time Hacksaw smiled in the interview. I had also asked him how he viewed what he had done and this was his reply:

> We were fighting a war against the Republican movement. They were killing the RUC [Royal Ulster Constabulary], they were killing the UDR [Ulster Defence Regiment], they were killing the British Army. We are British so we had to stand up and say 'Whoa, are we going to step back and let this happen?' We put the fear in them too, if you're going to do that then we're going to do that to you too. Only we'll do it harder.

I then questioned Hacksaw about the shooting of a Catholic man and his 10-year old son in a car on Cambrai Street in North Belfast (detailed in Chapter 4). When I asked 'Was that someone who was targeted?' he replied:

> Well, we missed the target that night and that was the next target that came. Instead of going home empty-handed, we took him out of the picture. We weren't really looking for this one. We were looking for another one who was in the IRA. And then this other guy came along.
>
> It was denied that this one was in the IRA, but the way I see it is that throughout the Troubles if the British Army or the RUC shoot somebody then they always say that the victims were killed in active service and that they are lieutenant colonels or rank or they're volunteers in the IRA but when the Loyalists hit one of them, and we've got a full dossier on them, it's denied, and I believe that it's all a cover-up to make the Loyalists look as if they are just hitting ordinary Catholics.
>
> Three of us went out to do the job, a driver and two gunmen. I was the driver; they both shot him. Two of us got done for it; one was never caught. We believed that once we came back from a shooting you weren't allowed to speak about it. It was loose talk then. If you were caught talking about it then you got it, so it had to be secret. If me and you and somebody else went out tonight, once we had done that mission and went back, the only ones who knew were the Commander and nobody else and that was as far as it went.
>
> You have to do what you're told, it's like if your father tells you to do something you have to do it, but it's more stricter (sic) than your father. It's hard to describe because if I don't do what I'm told then I'm putting your life in danger. And if I do something drastic while you're with me, the consequences could be we could be captured or else shot dead. The adrenaline is flying through you; you're really keyed up. Capture is the

> last thing on your mind when you are out on a mission like that. You see, I've a gun in my hand. If somebody tries to stop us we have to shoot our way out.

The power of language was at work again, turning Hacksaw and his two accomplices into the victims of the piece by stating that his group considered taking the chance of hijacking the victim's car and traversing Shankill with the guns in the car: 'We weren't going home empty-handed because if we had been caught we were getting big time.' He used the 'empty-handed' phrase as if the group had been on a fishing trip.

I then watched for some recognition of the absurdity of this phrase in his facial movement, but there was none. I listened to language with the power to transform a distance of a few hundred yards on the group's home turf into a much greater achievement: 'way up the Shankill with the guns in the car.' I watched Hacksaw smile when I asked how he felt about the shooting. 'It's sad,' he said.

That smile was most extraordinary. I have shown the film to students and they instantly identify it as the smile of a psychopath, relishing a murder, and providing an insincere apology about how sad he found a murder. But is this correct?

The late Stanley Milgram, an American social psychologist, commented on smiles in his research on obedience. Milgram conclusively demonstrated the power of a situation to produce obedience to authority (cf. Beattie 2018a). At Yale University in the 1960s, he conducted a number of psychological studies that involved assigning participants to the role of teacher. They were told to apply what appeared to be electric shocks to learners when the learners made mistakes in memory tasks. Milgram found that these ordinary students often continued to apply electric shocks to a degree that would have killed the learners if the shocks had been genuine. In fact, about 65% of the participants in his studies continued to obey the experimenter to the bitter end.

Milgram argued that the ability of participants to offset personal responsibility for their actions by indicating their actions were merely instrumental and responses to legitimate authority was crucial to such obedience. I noted that Hacksaw saw orders from members of his organisation as similar to directions from his father. Milgram also drew attention to the participants' uses of other strategies (called *psychic blinkers* by some psychologists) to minimise their guilt. The strategy is intended to 'shut out the awareness that the victim is a living, suffering fellow being.'

According to one of Milgram's participants, 'You really begin to forget that there's a guy out there, even though you can hear him. For a long time, I just concentrated on pressing the switches and reading the words.' Hacksaw talked about his victim not by name, but treated him as a member of two overlapping categories by describing him as 'a random' and 'a Catholic'.

Milgram also commented on the way that individuals often smiled when they responded to an authority figure's instructions to apply electric shocks. Again, the smiles represent some of the most memorable moments from the grainy black and white films of those infamous experiments. Milgram emphasised that such smiles should be seen as releases of tension rather than indications of enjoyment.

The interactional and functional natures of smiles

Here is another example of how smiles can be used in less obvious situations. A number of years ago, Vicky Lee and I attempted to understand the psychological benefits of emotional disclosure to determine what aspect of 'getting things off your chest' makes us psychologically and physically healthier (Lee and Beattie 1998; 2000). We wanted to analyse in detail what people said and how they said it when given the opportunity to talk to a confederate (who was one of the experimenters) about a negative event (the choice of event was entirely theirs as is often the case in emotional disclosures).

One of our participants described what happened whilst she was working as a care assistant and nurse's aide in a US hospital. She had been asked to transfer a patient from his wheelchair to his bed. However, what should have been a relatively simple and smooth operation did not go according to plan. The patient tried to walk without help, fell, cracked two ribs that subsequently punctured his lung, and died 2 weeks later. Again using the transcription method outlined in Chapter 4, we examine what the aide said (E = experimenter, S = subject):

1. S: ↑U:m (0.5) well the ↑negative event happened when I was (0.3) I was
2. working at the same hospital in America. (0.7)
3. Em (0.6) I was working as a- a care assistant and °a nurse's aid,° (0.5)
4. and er I was supposed to trans:fer a <u>pati</u>ent from his wheelchair into the
5. bed
6. E: yyeh
7. S: and erm (0.2) I didn't <u>ac</u>tually know that this- this patient had a ((sort
8. of history of not doing what he was <u>to</u>:ld.)) ((*smiley voice*))
9. E: yeh
10. S: So em (0.7) I put him in his chair by the side of the bed, (0.5)
11. and I went out to call the nurse, (0.2) 'cos the nurse had to be present
12. when you did a transfer
13. E: right
14. S: and er (0.2) as I sort of left his side (0.2) he got up and decided to try
15. and walk by him↑self:: (0.9)

16. So em (0.4) when he got u::p (0.6) he er he ↑fell,
17. he fell sideways into em like- like er a bar alongside the ↑wal1:
18. E: yeh
19. and err (0.2) he cracked two ribs. (1.0)
20. And er I- I TURNED ROUND JUST IN TIME TO SEE HIM
21. FA:LLING,
22. SO I- I MANAGED TO GET THERE BEFORE HE HIT THE
23. FLOOR,
24. AND I MANAGED TO SORT OF CUSHION HIM as he hit (0.2)
25. you know, got down,
26. E: right
27. S: and em (0.3) hit the emergency cord, and called the doctors and
28. everything.
29. But em (0.5) then we found that he had these two broken ribs,
30. and em (0.2) ↑three days later (0.2) it em punctured his ((lung (0.4)))
31. ((*laugh*)) and he DI::ED ((two weeks later)) ((*laugh*))
32. and em (0.2) I felt like it was all (0.3) my responsibility and my fault
33. for having left him
34. (0.3)
35. E: right
36. S: and er (0.2) we had to have a (0.1) disciplinary council to (0.4)
37. vin:dicate me and say that it wasn't my fault, (0.8)
38. and er (0.1) I just remember ((be:ing)) ((*smiley voice*)) re:ally really
39. down and really (0.2) feeling, (0.4)
40. apart from the fact that you know that I- I actually (0.6)
41. I felt guilty for the fact that I hadn't been there and- and like to
42. help him, (0.2)
43. E: right
44. (0.1)
45. S: even though it wasn't really my fault 'cos I turned round and (0.3)
46. you know he did it on his ↑ow↓n.
47. E: yeh
48. (0.4)
49. S: But I just remember feeling ab:solutely awful 'cos I felt
50. ((RESPO::NSIBLE)) ((*laugh*)) ((FOR HIM DYING)) ((*smiley voice*))
51. AND THEN I HAD ALL THIS ER (0.3) °palaver with the (0.1)
52. council, and had to go and give° (0.1) ((testi↑MONY:: and)) ((*smiley*
53. voice)) ((all the rest)) ((*laugh*)) of it.
54. So (0.2) it was a real nightmare (((1.2))) ((*laugh*)) ((basically)).
55. ((*smiley voice*))
56. Em (0.6) and- and I had to go and see (0.5) em the council ↑every
57. ((couple of weeks)) ((*smiley voice*)) 'cos (0.3) I got really tearful and
58. °upset about it,° and it er (0.1) shook me up a bit.

to cover amusement, *relief*, sensory pleasure, pride in achievement, the thrill of excitement, satisfaction and contentment'. He proposed that these different positive emotions could be considered members of the enjoyment family and noted that 'The thrill of excitement, relief, contentment, might all be different variations on a common theme, just as annoyance, fury, resentment, and outrage are all members of the anger family'. A research participant might well have experienced genuine relief whilst talking about some of the more negative aspects of an event.

The transcription of the account of the hospital aide described above revealed that she showed more Duchenne smiles and eye gaze whilst talking about the most negative parts of the trauma; she looked more at the researcher at those times.

Why did she exhibit more eye gaze? Argyle and Cook (1976: 91) argued that 'people who look more (and with longer glances) generally create a more favourable impression and are liked more'; see also Exline and Winters 1965 and Mehrabian and Friar 1969. Thus, the speaker could have used increased eye gaze as part of a self-constructive process to portray herself as a good, genuine and credible person. In Graham and Argyle's (1975) terms, eye gaze could have been used to strengthen this self-construction.

However, although Rutter and Stephenson (1979) did not question the relationship between gaze and affiliation, they argued that the *primary* function of eye gaze is to monitor information from others, rather than communicate affect. Thus, the increased levels of eye gaze during the most negative parts of the disclosure may be attributable to the speaker's desire to see how negative material is received. Emotional disclosure in the presence of another person is an interactive process, whether the person is a best friend, therapist or researcher. Clearly, eye gazes and smiles are vital parts of emotional disclosure.

The hospital aide gave a detailed account of an incident for which she might have been blamed: the death of a patient in her care. Her story, however, indicated she was not to blame. Her constructed account used a variety of means to assign responsibility to the patient. She did not necessarily 'get everything off her chest'. What she did instead was get a particular *account* off her chest. Duchenne smiles, presumably of relief, appeared when she recounted some of the worst bits of the story. When reporting those bits, she watched the researcher carefully to see how her story was received. Her actions may tell us something about the psychological mechanisms underpinning emotional disclosures.

What happens during emotional disclosure? Jamie Pennebaker (2000: 13) states that 'The net effect of constructing a good narrative is that our recollection of emotional events is efficient—we have a relatively short, compact story—and undoubtedly biased'. Pennebaker goes on to say:

> Ironically, then, good narratives can be beneficial in making our complex experiences simpler and more understandable but, at the same

> time, they distort our recollection of them. Translating distress into language ultimately allows us to forget or, perhaps a better phrase, move beyond the experience.

But this set of claims begs a lot of questions. What constitutes 'a good narrative'? What degree of bias can we tolerate? How is the degree of conflict in our memory related to how we construct an account and our ability to 'move beyond the experience'? Our case study and similar analyses (see Lee and Beattie 2000) demonstrated that issues explaining *why* events occurred are every bit as important as narrating *what* occurred. Any *coherent* narrative needs to contain these elements, and especially when the narrative is encoded into a 'short, compact, story' form.

However, it is neither efficient nor accurate to attempt to describe the process in terms of word counts of certain linguistic categories which is how Pennebaker himself has proceeded with his automated system LIWC (Linguistic Inquiry and Word Count). That, quite simply, is not how language works. The nurse's aide built her story around a patient who wilfully did what he liked in contrast to her need to follow rules. The incident was not a case of medical negligence, but a case of two differing individuals, with responsibility for the accident and subsequent death firmly attributed to the patient. The nonverbal behaviour of the speaker helped her construct herself as a decent, caring aide who did everything she could have done in the situation. But there may be other clues in her laughter and Duchenne smiles while describing the worst parts of the event that psychologists do not fully (or even partially) understand. Her eye gaze at these points in the disclosure narrative were aimed to determine how her account was received (positively in this case) and her Duchenne smiles may have been expressions of relief at devising an account that worked socially and in terms of attributional responsibilities (why the incident that led to the death of a patient occurred). She was not to blame for what occurred; that was the paramount point in her account.

Perhaps the most important lesson from this case study is the suggestion that emotional disclosure is not necessarily a straightforward cognitive or emotional process, as Pennebaker and others have so readily assumed—at least not primarily. Emotional disclosure is a social *interactional* process. If a person constructs an account of personal trauma that no one accepts, my guess is that the patient will not feel better physically or psychologically. This is, of course, an empirical issue that researchers may try to resolve in the future.

Critical to an individual's production of an emotional disclosure as discussed in this chapter is an individual's monitoring of the receipt of his or her message by her interlocutor (acting in a supportive manner to facilitate the disclosure process), and the genuine smiles of relief of the speaker as these critical components of the message are received and accepted. The social and the personal aspects overlap.

A critical issue for terrorists, violent doormen and others engaged in unequal contests is to devise versions of events that *they* (and others) can live with. Perhaps, I should have added that I interviewed Hacksaw in the garden of a house party at which he was a very welcome and popular guest. He was not living a lonely, isolated life in attempting to deal with the emotional consequences of his actions. His statements such as 'we missed the one we were looking for', 'they've denied he was a terrorist...but who really knows', and 'I was pleased when the 10-year-old boy pulled through' allow him to fit into the group easily. Psychopathic killers usually garner few social invitations when their history is known.

Similarly, the doormen devised versions of their events that helped them bond with one another and even with everyday punters who loved their stories from the door. I discussed this in a recent novel called *The Body's Little Secrets* (Beattie 2018b). The research participant who worked as a nurse's aide in the US gained sympathy, support and acceptance from her story. Smiles are not separate from all other aspects of communication; they are parts of the communication process. You may come to understand what smiles, even Duchenne smiles, mean as you watch them unfold alongside talk when terrorists and others relay their version of events.

Smiles and the shaping of memories

Of course, no talk accompanies the smile of a trophy hunter in a photograph but there are accompanying channels of nonverbal communication (positions of hunter and animal, distances, posture of hunter—all props for both the killer and the killed). So what possible interactional functions could smiles have in this context? They too could be laying down a version of the events, an account of what the hunt was all about—bravery (the huge animal laid out at the hunter's feet; no fear to mask with a non-Duchenne smile), professionalism – the clean kill (the wounds not exposed), the lack of negative emotion, the lack of any moral uncertainty (no nonverbal leakage here to indicate uncertainty).

But how is this interactional? After all, the only person in the immediate environment seeing this is the photographer. It is interactional for the viewers of the image, including the subject him or herself. It is interactional because it affects our cognitions. Images change our memories of events. We have known this since the classic research of Sir Frederic Bartlett (1932: 61) at the University of Cambridge. He began by recognising the *constructive* nature of human memory and devised ingenious experiments to show this constructive and transformational process in action. He noted that

'Remembering is rapidly affected by unwitting transformations: accurate recall is the exception and not the rule'.

When presented with ambiguous stimuli, Bartlett's subjects tried to make sense of them ('effort after meaning' in his words). He showed that a visual image to be understood in its most general sense of visualisation (and not just in terms of photographs) had the biggest effect on recall. The visual image also had 'the general effect of setting up an attitude of confidence which has nothing to do with objective accuracy'. In other words, a visual image has a major effect on what people remember and makes them more confident about its accuracy. Such images convince people that they remember information or events more clearly.

No human being can ever recall an event with perfect accuracy. Human memory is necessarily limited. That's why we take photographs: as reminders that keep our stories alive. However, photographs serve as more than reminders or illustrators; they play a major role in the reconstructive process. Garry and Gerrie (2005) demonstrated that you can even generate false memories in people using doctored photographs from their own lives (like a photograph of a person as a child and her parents in a hot air balloon). Garry and Gerrie said 'photographs might give subjects some kind of cognitive "springboard" allowing them to generate thoughts, feelings, details, images—the hallmarks of genuine memories—more easily than with verbal descriptions. Subjects confused these mental products for genuine experience, a process called source confusion' (p. 322).

Photographs are important in the stories we tell because they are set up to reveal, they are arranged, they conceal and distort. They display selected types of nonverbal communications and can change our memories of events as we add thoughts, feelings and details to the images. A photograph after a hunt preserves what we remember and also hyper-ritualises aspects of behaviour to define the relationship of the hunter and the hunted. These images are critical to the trophy hunting culture and are intended to attract admiration and praise for the photographic *versions* of the events.

The implications for analysing the smiles of trophy hunters should be clear. We cannot (immediately) infer specific satisfactions like achievement from the smiles alone. In the case of the traumatic incident involving a hospital aide and the death of a patient analysed above, Duchenne smiles were associated with the aide's apparent relief at disclosing her account of personal responsibility in an acceptable manner while she used eye gazes to monitor the interlocutor's receipt of the information.

Do the Duchenne smiles of the hunters signal relief at *winning a battle* with a large carnivore, relief that the photographer approves of the killing, or the anticipated pleasures of posting the image online and gaining admiration for killing a large beast? Or do the Duchenne smiles merely represent controlled expressions intended to preserve memories of an event? Without knowing onset and offset times, it is much harder to identify true spontaneous Duchenne smiles. Clearly these issues are worth exploring in more detail in the future.

In the meantime, we should be careful about labelling the smile of a trophy hunter in a photograph as that of a psychopath or even as menacing. Kendall Jones complained about such labelling in a 2014 interview with Bill McGrath. Such photographs are in many ways far more sinister because, as components of broader, highly configured images, they serve as tools to change the memories of the protagonists and attract the admiration of those with similar values. They propagate the trophy hunting culture by changing human memories of events. They constitute manufactured and crystallised moments in time that allow viewers' minds to fill in the details of the hunt with admiration and awe.

Like luxury cars and designer clothes, some of the rarest and most beautiful animals in the world become lifeless and soulless commodities used to enhance personal aggrandisement for certain types of people, The dead animals are arranged in carefully constructed tableaux to emphasise certain aspects of the killing and downplay others. Our attention is drawn to specific features in a certain sequence: the smile, the dead animal on the ground, then back to the smile.

In our research on climate change, we used eye tracking as experimental participants read climate change messages to show how attentional focus in the form of individual gaze fixations affected what participants' recalled from the messages (Beattie et al. 2017). Gaze fixations were affected by the psychological characteristics of the research participants. For example, optimistic individuals, who tended (quite literally) to look on the bright side of life (Isaacowitz 2006), were more likely than pessimists to focus on the more positive aspects of the messages. They tended to focus on arguments that climate change data was open to question rather than being scientifically clear and unambiguous and constituting very bad news that could have depressed their moods.

Optimists also recall and frame messages about climate change more optimistically because of their biased patterns of visual attention (Beattie 2018c). The pattern of fixations and perceptions provides the basis for the memory. The patterns differ for optimists and pessimists because of their desires to regulate their mood states linked to their underlying personalities (Beattie and McGuire 2020). The same bias applies to trophy hunters: 'Just look at that smile. I've never felt so proud. That was some achievement. That was some dangerous beast. I killed it clean. It never suffered.' The images are the record, the memory, the truth. They allow for self-aggrandisement of certain kinds of people and do it well. But what kinds of people exactly?

SUMMARY

1. Smiles, even Duchenne smiles, are more complex than we may first think (and many researchers seem to assume). A Duchenne smile displayed by a person talking about causing a death directly or indirectly does not *necessarily* indicate an underlying pathological personality.
2. Smiles are important aspects of nonverbal communications and fulfil a range of functions.
3. We considered smiles in emotional disclosure and the smiles of a terrorist discussing his actions to attempt to uncover their possible *interactional* functions—relief at getting something off their chest in a constructed account that is being accepted by the interlocuter.
4. We considered the smiles of trophy hunters while posing for photographs with their prey.
5. Smiles in photographs are also in a sense interactional, both for the viewers of the image, and the subject themselves – they influence others (and ourselves) by changing our memory for events.
6. Photographs are more than illustrators or reminders of events. They play a major role in the reconstructive processes of human memories.
7. We have known about the reconstructive nature of human memory since Sir Frederic Bartlett conducted research at the University of Cambridge in the 1930s. He devised ingenious experiments to show the reconstructive and transformational processes in action and wrote that 'Remembering is rapidly affected by unwitting transformations: accurate recall is the exception and not the rule'.
8. Trophy hunters use images to construct a visual account of what the hunt was all about – bravery (the huge animal laid out at the hunter's feet; no fear to mask with a non-Duchenne smile), professionalism – the clean kill (the wounds not exposed), the lack of negative emotion, the lack of any moral uncertainty (no nonverbal leakage here to indicate uncertainty).
9. Photographs are important in the stories we tell because they are set up, they are arranged, they conceal and distort. We display particular types of nonverbal communication in them. This changes our memory for the event as we add thoughts, feelings and details to the image later.
10. This affects what we remember about the hunt and, of course, it hyper-ritualises aspects of behaviour to define the relationship between the hunter and the hunted. These images are critical to the culture of trophy hunting and attract much admiration and praise for this *version* of events.

11. The smiles in trophy hunting photographs propagate the culture by changing human memories of the events. The photographs manufacture and crystallise moments in time and allow viewers to fill in details of hunts with admiration and awe.
12. I argued in this chapter that the smiles in such photographs are not necessarily 'those of the psychopath' or even 'menacing smiles'. In many ways, they are far more sinister in that they are components of broader, highly configured images and are instrumental in changing the memories of protagonists and attracting the admiration of those who hold similar values.
13. Such photographs propagate the trophy hunting culture by changing basic human memories.

Chapter 6

The personality of the trophy hunter

The hunter and his helpers.

> I do not like the killers, and the killing bravely and well crap. I do not like the bully boys, the Teddy Roosevelts, the Hemingways, the Ruarks. They are merely slightly more sophisticated versions of the New Jersey file clerks who swarm into the Adirondacks in the fall, in red cap, beard stubble and taut hero's grin, talking out of the side of their mouths, exuding fumes of bourbon, come to slay the ferocious white-tailed deer. It is the search for balls. A man should have one chance to bring something down. He should have his shot at something, a shining running something, and see

it come a-tumbling down, all mucus and steaming blood stench and gouted excrement, the eyes going dull during the final muscle spasms. And if he is, in all parts and purposes, a man, he will file that away as a part of his process of growth and life and eventual death. And if he is perpetually, hopelessly a boy, he will lust to go do it again, with a bigger beast.

John D. MacDonald
A Deadly Shade of Gold

John MacDonald is an American novelist who began as a short story writer of pulp detective novels for forty bucks a story. *A Deadly Shade of Gold* is the fifth novel in his Travis McGee collection. The plot revolves around a solid gold Aztec statue.

In this chapter we consider the evidence that certain personality types are associated with trophy hunting. The trophy hunters describe it as a noble activity with positive implications for conservation, MacDonald reminds us that not everyone shares this view. This chapter considers what science tells us about personality types drawn to the activity.

A number of psychologists have argued that individuals with certain types of personalities are *more likely* to be attracted to trophy hunting than others. The dimensions thought to be critical are those connected with *empathy*, *callousness* and *entitlement*. Indeed, some evidence suggests that individuals with high levels of *narcissism*, *psychopathy* and *Machiavellianism*—often cited in the psychological literature as the Dark Triad (Furnham et al. 2013)—may be especially prone to engaging in trophy hunting.

In terms of their underlying characteristics, 'narcissists have a grandiose sense of self and crave positive attention; Machiavellians manipulate social situations; and psychopaths are callous and lack empathy' (Meere and Egan 2017). Self-reported Dark Triad ratings are correlated with a number of antisocial behaviours including delinquency and aggression in children as well as risky and sensation-seeking activities (Crysel et al. 2013). According to Peter Jonason from the University of Western Australia and colleagues, the traits characteristic of the Dark Triad are thought to be associated with a 'compromised' or 'dysfunctional morality' (Campbell et al. 2009) in that they 'value "self" over "other" in a way that violates implicit communal sentiments' (Jonason and Webster 2012). Evidence collected by Phillip Kavanagh and his group from the University of South Australia suggests that 'individuals with higher levels of the Dark Triad demonstrated less positive attitudes towards animals and reported engaging in more acts of animal cruelty...These results suggest that those callous and manipulative behaviours and attitudes that have come to be associated with the Dark Triad are not just limited to human-to-human interactions, but are also consistent across other interactions' (Kavanagh et al. 2013: 666).

Whether results like this have *direct* implications for the personalities of trophy hunters depends, of course, on how you view trophy hunting. It is important to note here that there is no direct assessment of these personality traits in trophy hunters.

If you assume that you need a less positive attitude towards certain species of animals and additionally assume that trophy hunting is a (sanctioned) act of animal cruelty, it may seem reasonable to infer that these results are of some relevance to trophy hunters. As we have seen, trophy hunters would beg to differ about both the underlying attitude (admiration and love in their view) and about the cruelty of the acts although the acts may be reconstructed dramatically through the staging of the post-kill photographs and the reconstructive aspects of human memory. Direct research focussing on the personality characteristics of trophy hunters is still waiting to be carried out, but if you consider the dimensions of the Dark Triad in turn, they would seem to be of prima facie relevance to the activity of trophy hunting.

The rise and further rise of narcissism

Even a superficial consideration of trophy hunting websites would suggest that narcissism is likely to be a critical and obvious dimension of the activity. Narcissism is a personality trait 'associated with an inflated, grandiose self-concept and a lack of intimacy in interpersonal relationships' (Campbell et al. 2009). Narcissists are thought to suffer from 'extreme selfishness, with a grandiose view of their own talents.' In the most basic language, people who are highly narcissistic think that they are better than others in many dimensions, including their looks, intelligence, creativity, professions, and other areas but as Twenge and Campbell (2009: 19) noted, they are not:

> Measured objectively, narcissists are just like everyone else. Nevertheless, narcissists see themselves as fundamentally superior—they are special, entitled, and unique. Narcissists also lack emotionally warm, caring, and loving relationships with other people. This is a main difference between a narcissist and someone merely high in self-esteem: the high self-esteem person who's not narcissistic values relationships, but the narcissist does not. The result is a fundamentally imbalanced self—a grandiose, inflated self-image and a lack of deep connections to others.

Of course, therein lies the psychological 'rub'. If you see yourself as superior to others but are not actually superior based on objective indicators, how do you maintain your inflated level of self-esteem? The answer is that you must engage in a variety of strategies to maintain and develop your self-image.

For example, narcissists have a need to talk about their achievements to seek affirmation. Whenever and wherever possible, they need to broadcast their accomplishments (social media are ideal tools for this) to seek the maximum amount of affirmation. They focus on their physical appearance (amongst other attributes) and carefully select images that they present on social media or elsewhere. Selfies and photoshopping are important tools of narcissists. They value and display material goods, especially designer goods, to *instantly* communicate their superior social status (after all you only get to make a first impression once).

In social interactions, they try to make sure that conversations centre on them and they attempt to elicit compliments. Changes in appearance and wearing different clothes are necessary parts of their quests for compliments. In social relationships, narcissists seek out trophy partners who make them look good. Furthermore, since they lack warm and caring relationships, they often manipulate and exploit other individuals to ensure that they continue to look good relative to others.

The interesting and pertinent question is to what extent trophy hunting and the displays of dead lions and other large prey at the feet of hunters can be construed as part of a narcissistic strategy to elevate social status and maintain inflated levels of self-esteem. It is obvious that the postural arrangement of a hunter and a dead animal, especially one of the big five ('The lion is not called the *King of the Jungle* for nothing! What does that make me, buddy?'), is an iconic (and unconsciously understood) signal of elevated status vis-a-vis the animal. The message to be conveyed is that these feared beasts with their jaws held open by the hunter are nothing compared to the fierceness and bravery of the hunter. Additionally, trophy hunting is extremely expensive and intended to persuade a social media or other audience to admire the skills, courage and *wealth* of the hunter. Trophy hunting images are almost certainly designed to maintain a degree of narcissistic flow.

Narcissism is usually measured using the Narcissistic Personality Inventory (NPI) developed by Robert Raskin and Howard Terry at the Institute of Personality Assessment and Research at the University of California at Berkeley in 1988. The NPI uses pairs of 40 statements. In each case one choice reflects narcissistic tendencies, the other does not:

A: I prefer to blend in with the crowd.
B: I like to be the center of attention.

A: I can live my life any way I want to.
B: People can't always live their lives in terms of what they want.

A: I will never be satisfied until I get all that I deserve.
B: I will take my satisfactions as they come.

A: I am no better or no worse than most people.
B: I think I am a special person.

A: I try not to be a show-off.
B: I will usually show off if I get the chance.

A: I am much like everybody else.
B: I am an extraordinary person.

It is quite clear from this list that showing off ('I like to be the center of attention'), standing out from the crowd and being the focus of attention are core elements of narcissism. Trophy hunting obviously satisfies this aspect of personality. A sense of entitlement ('I can live my life any way I want to') that allows a narcissist to feel entitled to do anything he or she wants to do is also key. The concept of entitlement with regard to certain species of animals could be a key aspect of the narcissism of trophy hunters.

In their excellent book *The Narcissism Epidemic,* Twenge and Campbell (2009) discussed some common misunderstandings about narcissism. One common misunderstanding is that narcissists are really just people with high self-esteem. Twenge and Campbell points out that narcissists do indeed have high self-esteem (and the danger is that techniques used to increase self-esteem can also increase narcissism), but they argue that 'narcissism and self-esteem differ in an important way. Narcissists think that they are smarter, better looking, and more important than others, but not necessarily more moral, more caring, or more compassionate' (p. 24). In other words, they are only interested in being better than others in certain *key* aspects of life rather than all aspects of life.

Another common myth is that narcissists are fundamentally insecure individuals with low self-esteem and we should therefore pity or help them. Some psychologists have suggested that narcissists are just wearing a mask to conceal their low self-worth. Again, Twenge and Campbell argue that that is unlikely to be the case on the basis of research measuring the self-esteem of narcissists using the Implicit Association Test (IAT) discussed in Chapter 3. The IAT measures underlying (and unconscious) associations between concepts, in this case, the strength of the associations of the *me* and *not me* concepts with positive or negative words. The results suggest that narcissists respond more quickly when *me* is paired with good, wonderful, great and right, and slower when paired with bad, awful, terrible and wrong. In other words, narcissists unconsciously associate *me* with more positive attributes than do non-narcissists.

According to Twenge and Campbell, 'It turns out that deep down inside, narcissists think they're *awesome*' (2009: 27). Using the IAT, the evidence suggests that narcissists exhibit higher unconscious self-esteem than non-narcissists when given words like assertive, active, energetic, outspoken, dominant and enthusiastic (versus quiet, reserved, silent, withdrawn and inhibited). However, narcissists scored average on words like kind, friendly, generous, cooperative, pleasant and affectionate (versus mean, rude, stingy, quarrelsome, grouchy, and cruel). Twenge and Campbell conclude: 'Narcissists have very similar views of themselves on the inside *and* the outside. They are secure and positive that they are winners, but believe that caring about others isn't all that important' (p. 27).

Another technique that psychologists use to measure self-esteem is the name–letter task. Participants are asked to rate the letters of the alphabet according to how *beautiful* or *likeable* the letters are. Rating the letters in the participants' own names, especially the first letters, as more likeable or beautiful was seen as a powerful indicator of their inner self-esteem because it did not tap into conscious or deliberative processes. Again, the results suggest that narcissists think that the letters in their names are powerful and assertive, a little more beautiful, but not kinder or more nurturing. In other words, narcissists love many aspects of themselves, even the letters in their own names.

Twenge and Campbell also examined whether narcissists are in fact better looking than the rest of us. When you show people photographs of narcissists and non-narcissists and ask them to rate the subjects' appearances, it turns out that narcissists are not actually more attractive. However, they are very careful to select flattering pictures of themselves to post on social media and have enough pictures taken to ensure that some are flattering. This has obvious implications for trophy hunting shots because an important requirement of the shots is to flatter the courageous, good looking, and well-resourced hunter. Clearly, the images that are posted will be selected very carefully.

Narcissism is a major societal issue because it is increasing dramatically as evidenced by the title of Twenge and Campbell's book: *The Narcissism Epidemic* (2009). They point out that almost every trait related to narcissism (assertiveness, dominance, extroversion, self-esteem and individual focus) became far more common between the 1950s and 1990s. Twenge and Campbell noted that two-thirds of US high school students said they expected to be in the top 20% of performers in their jobs and stated that 'In a recent study, 39% of American eighth graders were confident of their math skills, compared to only 6% of Korean eighth graders. The Koreans, however, far exceeded the US students' actual performance on math tests. We're not number one, but we're number one in *thinking* we are number one'. Additionally, in the US, 'the ethnic group with the lowest self-esteem, Asian-Americans, achieved the *highest* academic performance' (p. 48).

The authors argue that one significant reason for this change in level of narcissism relates to major socio-cultural changes in society. At one point, clear social taboos exerted some control over people who were over-confident or appeared to love themselves too much. 'Who do you think you are?' was an oft repeated response when taboos were violated. Children were warned about feelings of entitlement and were told not to "show off'.

Now people are encouraged to love themselves and let others know about it, for example by wearing 'Diva" or 'Number One' tee shirts. Children are told to 'aim for the sky', 'be the best you can', and 'be number one'. Twenge and Campbell argue that changes in parenting practices are a major factor here. They also identify celebrities and the media as 'super spreaders' of narcissism. Television, and reality television in particular, present major opportunities for narcissists to harvest acclaim and for narcissism to be normalised (Kendall Jones springs to mind here; see Chapters 1, 3, 4, and 5 and next section).

Twenge and Campbell write that 'Narcissists are masters at staying in the spotlight; they love attention and will do almost anything to get it'. Narcissists thrive on public performances. Unlike many people who find appearing in front of a crowd extremely anxiety-provoking, narcissists love it. 'With the advent of reality TV, nonstop celebrity coverage, and instant fame, more and more narcissistic people are spreading their disease far and wide' (p. 91).

One side effect of the spread of narcissism is an increase of desire for fame. The authors cite a 2006 study reporting that '51% of 18- to 25-year-olds said that *becoming famous* was an important goal of their generation'. Only one-fifth of that number cited *becoming more spiritual* as an important goal (p. 93). A similar poll in the UK asked children what *the very best thing in the world* was. The most common answer was 'being a celebrity'. For those who don't achieve celebrity even in reality television shows, social media posts can make them celebrities within their social groups and allow them to increase (and anxiously measure) their narcissistic flow by the number of 'likes' they garner.

Social media allow users to present themselves in particular ways using necessary text, photos, and other necessary props. Blogs, YouTube, Facebook and other media allow people to design their own images to attract the kind of admiration they want. If admiration and feeling superior to others are of great importance (critical to your concept of self) and you don't necessarily care much about other people or other values, you can see immediately how trophy hunting can be used to maintain your narcissistic flow.

Narcissism, trophy hunting and language

Kendall Jones is an interesting case study here. At the time of one of her interviews discussed earlier, she was a 19-year-old cheerleader and trophy hunter studying at Texas Tech University who liked to pose in photographs with lions and cheetahs she killed with a crossbow. She has come in for a good deal of criticism, indeed severe (and sometimes violent) criticism from animal conservation groups for posted images that show her in a relaxed posture, heavily made up, with a natural looking Duchenne smile, with a clear dominion over the dead beast (eyes closed) at her feet. The images generate admiring comments from other hunters. It is the enigma in socio-biological terms of the female hunter.

Kendall agreed to the interview to explain why she hunts, to put the record straight and to give us an insight into her psychological motivations that may well have puzzled the Maasai of Eastern Kenya and other indigenous hunters.

But, as always with language, it is necessary to see how it is being used to construct both the activities themselves, and the debate to determine how the language is functioning. When you carry out even a cursory analysis of this discourse, you can see quite clearly that narcissism pervades the discourse itself, its function is to make her look (and feel) special in contrast to her critics.

She said 'Yes, I am a hunter but I'm a conservationist first and foremost. People like to jump to conclusions and most of the time it isn't their fault. Most people don't understand that hunting is and always will be a useful tool for conservation.' Not only does she draw on the familiar (and heavily criticised) argument that hunting and killing these animals can be connected to conservation, she indicates that her critics cannot recognise this connection because they are unable to overcome their intellectual inferiority.

During the interview, she applied the same argument to people's inability to interpret her facial expression:

> I understand that the photos might be a little unnerving to the inexperienced hunter or misinformed public, but they are simply documentation of my experiences. My 'menacing smile' as CNN called it, is not because I am happy about the death of a magnificent creature, but because I know that I am contributing to the betterment of the species and that I am providing much needed funding for conservation as a whole. I am actually doing something for the overall benefit of the animals I care about and I'm proud of that.

Kendall is not simply trying to take the moral high ground in her statements. She is also indicating that those who disagree with her (and critique her argument that hunting is linked to conservation) are unable to process and interpret correctly something as fundamental as nonverbal communication (including the kind of smile she shows in the photo). As noted in an earlier chapter, the smiles of trophy hunters may be a bit more psychologically complicated than this but I am not fully convinced that she understands the antecedents of her own smiles in trophy photos.

She asserts that her smile does not indicate menace or achievement. Rather it is benevolent and benign because she is doing something good for the African continent and for animal species. The smile, she explains, says something about her attitude towards the world which is a little bit hard to agree with when you view the image.

In the same interview she says that 'Hunting isn't a sport for everyone and I acknowledge that' and again she builds the argument that she likes 'hunting because of the challenge and mental strength needed to track and engage the animal...it is dangerous and difficult. Hunting is no easy feat. It takes a lot of physical strength and endurance to be a hunter. It also takes a lot of patience and skill to operate a rifle and even more to operate a bow. It's a very connected and even spiritual experience and you learn a lot from yourself.'

So this is not just saying that there is a diversity of experience or opinion, she is really saying that many people in the world, indeed, most people in the world wouldn't have the skills or attributes to be a successful hunter or to engage in this spiritual experience. In other words, her discourse around the narcissistic images of the dead animals is yet another strategy for building a gap between herself and others. It is about using self-construction for self-glorification to portray herself as a special, indeed spiritual, individual.

What is interesting about trophy hunters from a psychological perspective is that they don't just use images of themselves with their killed animals as semiotic devices to mark them as special. They also use the resulting 'setting the record straight' discourses to satisfy these same narcissistic drives.

One interesting and slightly startling aspect of the interview was Kendall's statement that 'I am a huge fan of *The Walking Dead*. If that ever happened in real life, I'd feel confident in my abilities to survive!' This is a very odd way of justifying killing animals in that if a fictional scenario ever happened then she feels confident that she could deal with it. So in other words, she is justifying killing animals by saying that if impossible scenarios ever happened then trophy hunting would have left her well equipped to deal with them. That may be another aspect of narcissism. The narcissist feels so special that he or she is capable of dealing with any situation, including scenarios that could never possibly happen. The self has been inflated to the realm of fantasy without a second thought.

Trophy hunting and the Dark Triad of personality

Some psychologists suggested that trophy hunting is associated with other personality dimensions alongside narcissism. For example, Xanthe Mallett, a forensic psychologist, in *The Conversation* (2015) suggested a close link between a number of socially aversive personality dimensions (namely the Dark Triad) and the hurting of animals which (one has to say) is a necessary part of trophy hunting. She cites the well known and highly publicised case of Walter Palmer, a US dentist who killed Cecil the lion near the Hwange National Park in Zimbabwe as evidence of suffering of animals. Such suffering is rarely mentioned by trophy hunters or disguised in the thousands of hunting images posted online, as noted earlier.

Pendergrass, Payne and Buretz (2016) point out that Cecil was not actually killed on Hwange National Park land; that would have been illegal. Cecil was lured across the park border to a private landowner's ranch. Palmer paid

US$50,000 for the privilege of killing Cecil and set off an immediate and global backlash. Mallett and others have drawn attention to the suffering of the animal. Cecil was initially only wounded with a crossbow and finally shot dead two days later before being beheaded and skinned. Cecil, however, was not an ordinary lion. He was a very famous lion—a major tourist attraction in Hwange National Park. Pendergrass et al. (2016) wrote that Cecil was 'renowned for being friendly towards tourists and was instantly recognisable because of his large size and distinctive black mane'. Cecil was also wearing a GPS collar 'which should have been apparent to all participants' (p. 69).

Mallett (2015) reminds us that intentional hurting of animals is an element of the standard test used to diagnose psychopathy and writes:

> Since the 1970s, research has shown that the majority of adults who commit violent crimes have a history of animal cruelty in childhood. Some studies suggest that up to 70% of the most serious and violent offenders in prison have repeated and severe episodes of animal abuse in their history.

Mallett also cites the work of forensic psychiatrist John MacDonald, who, in a well-known article in the *American Journal of Psychiatry* (1963) identified a connection between violence and the three dimensions of personality that constitute the Dark Triad: (1) Machiavellianism (manipulative personality, which, of course, gets its name from Machiavelli's book on the topic; see Christie and Geis 1970), (2) subclinical or 'normal' narcissism (with feelings of grandiosity, entitlement, dominance and superiority) and (3) clinical or 'normal' psychopathy (with characteristics such as high impulsivity and thrill-seeking combined with low empathy and anxiety; see Hare 1985).

Paulhus and Williams (2002: 557) analysed the relationship between these three personality dimensions in a non-forensic, non-pathological, high-achievement population (a sample of university students). They point out that all three dimensions share a number of important features: 'To varying degrees, all three entail a socially malevolent character with behaviour tendencies toward self-promotion, emotional coldness, duplicity and aggressiveness'. They also say that in the *clinical* literature, the connections between the dimensions had been known for some time (Hart and Hare 1998).

However, newer research demonstrates strong degrees of overlap in *non-clinical* samples as well, particularly between Machiavellianism and psychopathy (Fehr, Samson and Paulhus 1992), narcissism and psychopathy (Gustafson and Ritzer 1995) and Machiavellianism and narcissism (McHoskey 1995). Paulhus and Williams (2002) tested the interrelationships of these personality dimensions in their sample, and also tested their connections with other measures of personality (intelligence, overconfidence in performing various cognitive tasks and other factors).

They found that all three dimensions were correlated (with the strongest correlation between narcissism and non-clinical psychopathy), and that the dimensions were related but were not equivalent. The one commonality across the triad was *low agreeableness*.

The researchers also noted significant differences amongst the three dimensions, for example, only psychopaths (but not narcissists nor Machiavellians) exhibited low anxiety and were resistant to punishment; this is consistent with the general clinical view. They also found that narcissists and (to a lesser extent) psychopaths exhibited the most self-enhancement in various cognitive tests, whereas Machiavellians showed few signs of it. Machiavellians are, the researchers concluded, more reality-based in their sense of self. Narcissists exhibited strong self-deceptive components (with low insight).

Paulhus and Williams (2002) point out that the grandiosity and poor insight exhibited by narcissists have also been noted in clinical psychopaths (Hart and Hare 1998). They also found that the measure of non-clinical psychopathy was the best predictor of both self-report and behavioural measures of antisocial behaviour. Paulhus and Williams concluded that the Dark Triad of personality characteristics were 'overlapping but distinct' constructs, each dimension in the triad presenting with its own particular problems.

Kavanagh, Signal and Taylor (2013) analysed the relationship between the Dark Triad and aggression and criminality and point out that 'animal cruelty' is a "red flag" indicator for the propensity to engage in violent antisocial behaviours including intimate partner abuse (Volant et al. 2008), intra-familial violence (Khan and Cooke 2008), sexual assault (Simons, Wurtele and Durham 2008), and bullying (Gullone and Robertson 2008: 667). They tested 261 participants on standardised measures of psychopathy (Self-Report Psychopathy Scale-III), an adapted version of the NPI-16 to measure narcissism, and MACH-IV to measure Machiavellianism. They found (again) that all of the three measures were correlated, so they averaged them to form a composite Dark Triad score with higher scores indicating higher levels of this composite trait.

Next they assessed attitudes towards animals using a 26-item Attitudes towards the Treatment of Animals scale on which participants had to indicate the extent to which they were bothered by certain acts towards animals, including 'intentionally killing a wild animal while hunting'. The participants were also asked about behaviours in which they engaged: 'Have you ever intentionally killed an animal that was owned by yourself or by someone else for no good reason?' and 'Have you ever intentionally killed a stray, feral, or wild animal for no good reason?'

The analysis of Kavanagh et al. (2013) revealed that less positive attitudes towards animals were associated with higher levels of narcissism, higher levels of Machiavellianism and higher levels of psychopathy. In addition, the composite Dark Triad score was also correlated with attitudes to animals. Higher levels of psychopathy were associated with actual behavioural (and not just attitudinal) measures i.e. 'having intentionally killed a stray or wild animal for no good reason' and 'having intentionally hurt or tortured an animal for the purpose of teasing it or causing pain'. They say their results were unique but not surprising 'given the typical profile of those high on the Dark Triad' (p. 668).

Kavanagh et al. then attempted to determine the common element or elements in terms of these inter-locking personality dimensions. Their conclusion was that 'Callousness lies at the heart of the "dark core" of the "Dark Triad".' Those who kill animals for no good reason in this way and have a poor attitude to animals seem to suffer from a degree of callousness generally. Somewhat worryingly, the researchers also found that age was also a significant variable; younger people displayed higher levels of those traits than older people.

Jonason et al. (2013) suggested that empathy, or rather the lack of empathy, is another significant feature linking the three dimensions. Empathy plays a major role in the identification of psychopathy; indeed it is one of the two core dimensions. It also feeds into Machiavellianism but it has a more complex relationship with narcissism. Jonason et al. argue that a core route for women in relation to the Dark Triad may be narcissism rather than psychopathy or Machiavellianism.

Using a complex pictorial analysis of the data, Jonason et al. concluded that 'Moderation tests suggest the link between the Dark Triad and limited empathy might primarily be through narcissism in women but psychopathy in men' (p. 574). Furthermore, 'men who are high on psychopathy and thus have limited empathy may enact a risky lifestyle whereas women who are high on narcissism may enact parasitic relationship styles' (Jonason and Schmitt 2012). In conclusion, callousness and lack of empathy seem to be the key attributes of the Dark Triad and entitlement underpins narcissism which is another key attribute.

It is worth emphasising that Kavanagh et al. did not specifically focus on trophy hunters in their research which is something that the forensic psychologist Mallett mentions as she concludes her commentary piece in *The Conversation*. She wrote 'The problem is that understanding why people hunt for pleasure would require in-depth psychological assessments of a large number of hunters against evaluative measures for a whole range of personality traits, before we could try to figure out what people are feeling and what their motivations are.'

In other words, the Kavanagh study produced evidence that certain personality characteristics are associated with animal cruelty but, it is important to emphasise again that this research did not focus directly on trophy hunting per se. However, you can see that there is likely to be a close link between these personality dimensions and trophy hunting, given that trophy hunting necessarily involves animal suffering and therefore animal cruelty—despite the many reports of hunters and images portraying animal carcasses that suggest that most kills are 'clean'.

Unfortunately for the trophy hunting community, much of the documentary evidence (and common sense) suggest that considerable animal suffering is involved. A lack of empathy and a degree of callousness may well facilitate trophy hunting and may even be necessary personality traits for pursuing it. Trophy hunting and its depiction in photographs and films may well facilitate the maintenance of narcissistic flow (another necessary condition). This combination of personality characteristics may be lethal—and certainly is for the hunted animals. However, as is often the case, there is the proviso that more empirical work in this area is urgently needed, particularly more direct research on trophy hunters' personalities.

This research is clearly a move away from the evolutionary perspective, which might suggest that trophy hunting is somehow *the way things are and the way things were always meant to be*. However, as we saw in Chapter 2, the killing of large game cannot be explained easily in terms of food resources for families and the propagation of species in straightforward evolutionary terms. It was always more complex than that.

Now we have a term for those whose self-identity depends upon the constant admiration of others (narcissism). We have ways of measuring callousness and lack of empathy in the dimensions of non-clinical psychopathy and Machiavellianism. These developments, you could say, represent cultural evolution at work and we may be getting somewhere in our understanding of the psychology of trophy hunting. Trophy hunting is clearly not for everyone. The recent research on specific personality traits outlined here would seem to confirm that.

SUMMARY

1. It has been suggested that individuals with certain types of personalities are *more likely* to be attracted to trophy hunting than others.
2. The dimensions considered critical relate to empathy, callousness and entitlement. Indeed, some evidence suggest individuals with high levels of *narcissism*, *psychopathy* and *Machiavellianism* (the Dark Triad) may be especially prone to engaging in trophy hunting.
3. Self-reported Dark Triad ratings are correlated with a number of antisocial behaviours including delinquency and aggression in children, as well as risky and sensation-seeking activities.
4. Evidence suggests that 'individuals with higher Dark Triad levels demonstrated less positive attitudes towards animals and reported engaging in more acts of animal cruelty.'
5. Whether results like this have direct implications for the personalities of trophy hunters depends, of course, on each individual's view of trophy hunting. If you assume that you need

a less positive attitude towards certain animal species and also assume that trophy hunting is a sanctioned act of animal cruelty, it may seem reasonable to infer that these results are relevant to trophy hunters.

6. Trophy hunters, of course, would beg to differ (as we have seen) both on the underlying attitude (of admiration and love, they would contend) and on the cruelty of the acts (although the acts may have to be reconstructed by staging of post-kill photographs and utilising the reconstructive aspects of human memory).
7. Direct research on the personality characteristics of trophy hunters is still waiting to be carried out.

Chapter 7

Concluding remarks

The victor and the vanquished (the rifle with the telescopic sight gently leaning on the dead beast).

He was an old lion, prepared from birth to lose his life rather than to leave it. But he had the dignity of all free creatures, and so he was allowed his moment. It was hardly a glorious moment.

The two men who shot him were indifferent as men go, or perhaps they were less than that. At least they shot him without killing him, and then turned the unconscionable eye of a camera upon his agony. It was a small, a stupid, but a callous crime.

Beryl Markham
West with the Night

Beryl Markham was a British writer born in Rutland, England. She moved with her family to Kenya which was then colonial British East Africa. *West With The Night* was her memoir of her life in Kenya.

It seems that Ernest Hemingway was a great fan of her writing but not necessarily a great fan of the woman herself. Hemingway wrote 'but this girl who is to my knowledge very unpleasant and we might even say a high-grade bitch, can write rings around all of us who consider ourselves as writers...it really is a bloody wonderful book.'

Her description of the shooting of the lion is poignant, and sad, and telling. It is, of course, a highly personal perspective. What we have tried to consider in this book are the broader psychological considerations that may influence how we perceive trophy hunting as we try to understand why it flourishes in these times.

As I said at the beginning of this book, there has been a great deal of highly charged emotional debate and considerable outrage about trophy hunting. Many of the arguments concern the ethics of the activity and, even if we adopt a consequentialist philosophical position and argue that the fees from trophy hunting can and do feed into the local communities in desperate need of such income (the *ends*), whether the killing (the *means*) may ever be justified.

Clearly, the act of paying very large (and, on occasion, astronomical) sums of money to travel to Africa to kill trophy animals, particularly endangered and charismatic species, in the most rigged and unequal 'sport' evokes powerful emotions. The emotions connect to the wanton destruction, the taking of life, the destruction of beauty, the rigging in favour of the hunter, the boasting about it, and the endless, smiling images that stick in the gullets of many. The emotional response to trophy hunting is clearly compounded and amplified by the carefully choreographed display of carcasses in images which include the smiling hunter and the means of killing (that inflict considerable suffering on the animals).

And critically, the looks of the hunters add to this emotionally-charged response. The hunters are clearly self-satisfied with their achievements, boastful, pleased with themselves, proud; as pleased and as 'proud as Punch' one might say. The looks alone can lead to intense, inflamed anger.

Samuel Pepys recorded in his dairy that he took his wife to see *Pulcinella* in 1666, and as Mr Punch squawked about his wife beating and violence ('That's the way to do it!'). Pepys noted that he and his wife immensely enjoyed the comically gruesome spectacle of the gleeful, monstrous wife killer and baby killer and considered Punch a 'jolly good fellow', squawking about his achievements in this absurdist comedy.

Trophy hunters appear 'as pleased as Punch' in the real (too real) images of dead animals, killed to attain some unclear psychological benefit. 'Hardly sport', their opponents say and as absurd as Mr Punch crowing about his achievements with his slap stick.

This book has attempted to take a different angle on this issue and explored the psychology of the trophy hunter in several ways. It considered a number of psychological perspectives, starting with the evolutionary view and reaching the conclusion that even in evolutionary terms, the killing of large prey was always more than just about securing a large food resource for family. Hunting served as a social signal of *inclusive fitness* to a social group.

In the contemporary world, a dead animal is not necessarily a signal of courage, skill, or cunning. It represents material wealth and thus still has relevance to socio-biological arguments. However, the evolutionary argument left many questions unanswered. Why are some men drawn to trophy hunting rather than being contented with, for example, a sports car or a trophy girl friend? How does the evolutionary argument apply to female trophy hunters? In socio-biology, the signalling of resource is thought to be primarily the domain of men to attract mates. Women, it is argued, have other means to attract mates. But female trophy hunters have become major social and cultural phenomena that cannot be ignored.

Then we attempted to understand the psychological motivations of trophy hunters by considering studies that analysed posts on hunting websites. Some researchers argued that a number of apparently distinct psychological motivations for hunting emerge from these accounts. They identified the main *satisfactions* that hunters derive from this pastime. First and foremost is *achievement* (feelings of satisfaction related to performance). Secondary satisfactions are *appreciation* (enjoyment of the experience) and *affiliation* (strengthening of personal relationships and/or enjoyment of the company of others).

I suggested that a different approach might be needed here and proposed a more functional, top-down approach to the analysis of these accounts to uncover the motivations more effectively. I also warned about accepting any explanations of personal behaviours too readily. If psychology has shown us anything over the past few years, it is that many of our behaviours have unconscious and hidden origins. I looked outside trophy hunting to climate change and racial prejudice to support this view.

Some researchers tried to move beyond these verbal descriptions to identify the psychological satisfactions associated with hunting by analysing the nonverbal communications of hunters when they pose for photographs with their dead animals. Since some evolutionary anthropologists suggested that displays of dead animals have served as important elements in competitive displays throughout the evolution of hunter–gatherer culture and thus the psychological motivations to engage in this form of behaviour may be more implicit anyway and far less conscious.

From this perspective, verbally reported motivations may actually be incidental or of little consequence, although at times hunters can draw upon such discourse to explain their own behaviours ('men were born to hunt big prey and show what they're capable of'). The research discussed in this book revealed that certain types of nonverbal communications, particularly *spontaneous* and *genuine* Duchenne smiles, were significantly more common when hunters were photographed with prey than without prey and also when the hunters posed with large prey compared with small prey. Older hunters showed more Duchenne smiles than younger hunters. When posing with carnivores rather than herbivores, older hunters were significantly more likely to show Duchenne smiles. The researchers concluded that their studies produced 'independent evidence that displaying prey evoked satisfaction in some achievement context. Moreover, that old hunters actually show more satisfaction displaying large/dangerous prey than when posing with small/herbivorous prey suggests achievement-oriented satisfaction has not decreased with age' (Child and Darimont 2015: 9).

In this context, I argued that the Duchenne smiles were not necessarily natural nonverbal signs of achievement but on occasion represented hyper-ritualised displays ('Just think of what we've managed to bag here and what the folks back home are going to think about this. Wait till they see this!') selected specifically for posting on social media. The mouths of dead (and formerly ferocious) animals in trophy hunters' photos are sometimes propped open to show their teeth. The images convey psychological and political messages about power and dominion over all living things and about the joy of exercising that power—doing what comes naturally to men.

And what about women? What are they doing in these 'natural' celebrations of evolution? They appear to have taken the concepts of function ranking and subordination as described by the late Erving Goffman, and subverted them. The images propagate the basic political and cultural message of trophy hunting, about power and privilege and entitlement without moral uncertainties. The great animal species have become props in a story about the individual hunter and his or her great achievements, their values, their rugged individualism, their pursuit of freedom and resources. These are images intended to be admired. But the question then arises as to how trophy hunters can possibly justify this whole process.

In Chapter 4, I considered the processes of justification, introducing material from study areas outside trophy hunting to see how these processes operated. I argued that justifications provide us with structures for our narratives and stories. They give meaning to our lives. They emerge freely as we talk about what we've experienced and done. Common elements emerged in the accounts of hard men working on the doors of clubs, terrorists on the streets in Northern Ireland, and trophy hunters out in Africa. I suggested that the common elements include a logical calculus of attribution theory employed to objectify the assignment of causation and blame, the construction of self and victim, and the downplaying of the injuries inflicted by the protagonists. I noted that among several thousand photographs of dead trophy animals, very few showed their wounds.

We are creatures of language. It makes us human and separates us from all other species. It enables us to explain our actions. Without these justifications we would be lost.

I reminded myself that although Hacksaw, one of the terrorists I interviewed, found religion in prison, he nevertheless attempted to justify a murder he committed in a complex and insidious manner. But we need to see these things for what they really are—justifications. We use them habitually in our everyday descriptions, often with (seemingly) little awareness of what we are doing.

The *whole* discourse of trophy hunting needs to be considered to determine how these pervasive justificatory techniques are employed. Justifications are not used only to persuade other people about how and why events occur. They seemed to work well for Hacksaw, the doormen and the trophy hunters themselves. I argued that language, of course, helps us communicate with others, but it also helps us *privately* give meanings to our own lives. I suggested that these processes are intimately connected. This is how trophy hunters understand their lives and actions.

In Chapter 5, I returned to the issue of the beaming smiles of trophy hunters posing with animals they killed. I suggested a functional analysis of these images (and therefore of the smiles)—they affect what we remember. I am reminded of that old saying that 'There is no such thing as history, there are just versions of history'.

Photographs are important in the stories we tell because they are vivid; they remain with us after other aspects of the memory are forgotten. They are also set up, they are arranged, they conceal and distort. We display particular types of nonverbal communication in them. This changes our memory for the event as we add thoughts, feelings and details to the image. This affects what we remember about the hunt and the kill, and of course, it hyper-ritualises aspects of behaviour to define the relationship between the hunter and their prey. These images are critical to the culture of trophy hunting and attract much admiration and praise for their *version* of events – the trophy hunters' version.

The implications of analysing the smiles of trophy hunters should be clear. We cannot immediately infer specific satisfactions like achievement from smiles. A smile may indicate other emotions. Do the Duchenne smiles of hunters demonstrate their relief at having *won a battle* with a large carnivore? Do they indicate relief that the photographer appears to approve of the killing? Do the smiles portray the anticipated pleasure of posting the image online (and the subsequent joy of being admired for killing a large beast)? Or are the Duchenne smiles simply controlled expressions that constitute parts of the memories of events? I suggested earlier in this book that it is very difficult to identify genuine spontaneous Duchenne smiles without knowing the onset and offset times. These are clearly issues to be explored in more detail in the future.

In the meantime, we should be careful about labelling hunters' smiles in photographs as 'those of the psychopath' or even as 'menacing smiles'. They are (in many ways) much more sinister than that because they, as part of a broader, highly configured image, are instrumental tools. They are tools to change the memory of the protagonist and tools to draw in the admiration of those with similar values. They propagate the culture by changing human memory. They are, I argued, a crystallised and manufactured moment in time which allow our minds to fill in the details of the hunt with admiration and awe. Like luxury cars and designer clothes, some of the rarest and most beautiful animals in the world become lifeless, soulless commodities to personal aggrandisement for certain types of people arranged in a carefully-constructed tableau to emphasise certain aspects of the killing and downplay others. Our attention is drawn to certain features—the smile, the dead animal on the ground, the proud hunter beaming away, there to be envied (that, at least, is their intention). The images become the record, the memory, the truth. They allow for this self-aggrandisement of certain kinds of people, and these images do it well.

But what kinds of people? In Chapter 6, I considered the psychological evidence that certain personality characteristics, namely the Dark Triad of Machiavellianism, non-clinical narcissism and non-clinical psychopathy, are associated with animal cruelty. However, I pointed out, this research has not focused directly on trophy hunting per se, although you can see that there is likely to be a very close link given the clear documentary evidence, which suggests that considerable animal suffering is involved in trophy hunting. A lack of empathy and a degree of callousness (characteristic of the Dark Triad) may well facilitate trophy hunting (and may even be necessary conditions for trophy hunting), and trophy hunting and its depiction in images and films may well facilitate the maintenance of narcissistic flow (another necessary condition). This combination of personality characteristic may be indeed critical. But as is often the case, there is a proviso, more empirical work in this area is urgently needed, particularly more direct personality research on trophy hunters themselves.

Indeed, more original empirical work in psychology needs to be done more generally in this area especially in the sort of way that has been attempted here, which considers multiple perspectives on this same complicated and (from some points of view) puzzling issue. Only by furthering our understanding of how justificatory strategies percolate everyday descriptions of blameworthy actions (and why these work not just for others but for the benefit of the protagonists themselves), of how images in photographs affect what we remember and what we forget about the 'clean' deaths of our prey without suffering, of how nonverbal communication affects our evaluations of a scene and our mastery of the natural world, of how smiles 'leak' or mask inner states, of how personality affects our feelings of entitlement and our feelings of callousness towards other living things, will we ever get to grips with why some people feel this desperate psychological need to kill highly-prized, indeed precious, animals for their own selfish benefits.

We may also need to develop a neuroscientific understanding of these processes and determine why trophy hunting continues to flourish. I am reminded of what Elbert, Weierstall and Schauer wrote in the *European Archives of Psychiatry: Clinical Neuroscience* in 2010:

> The hunting behaviour in male hominids—who evolved from vegetarian ancestors—has developed since the Pliocene, i.e., since several million years. Being rewarded with social and eventually reproductive success led to hunting for ever bigger trophies. In the course of evolutionary adaptation, reward systems became reorganised to experience hunting behaviour positively...Hunting for men, more rarely for women, is... emotionally arousing with the parallel release of testosterone, serotonin and endorphins, which can produce feelings of euphoria and alleviate pain. (pp. 101–102).

Hunting (not just hunting of large prey) clearly changed us as a species and affected the neuronal circuitry of the human brain. Many life events (stress, maternal deprivation, drug addiction, gambling) affect us far more quickly than evolution. The fact that an activity is powerful enough to influence us over aeons at multiple levels including the physiological (through the alteration of neuronal circuitry) does not mean that we should not try to change it. I am struck by the words of Elbert and his colleagues: 'Hunting... can produce feelings of euphoria and alleviate pain.' As conscious, and sentient creatures, humans do not have to alleviate pain by inflicting it on other creatures to satisfy some evolutionary prerogative, nor indeed any other prerogative based around individual psychological functioning, semiotic signalling or group psychology.

Psychology may well hold the key to understanding trophy hunting, why it continues, why it flourishes in this narcissistic age of ours, and ultimately what can be done to combat it (if that becomes the goal). Human beings are, after all, capable of self-reflection (with the right prompts) and can develop insight into their actions. *Sometimes* awareness may be enough to promote some degree of change. But, of course, change also depends on other factors that hold a behaviour in place. We need to identify, analyse and deconstruct those factors and the processes of identification and analysis at *multiple* levels has only really just begun.

References

Allport, G. 1935. Attitudes. In C. Murchison (Ed.), *Handbook of Social Psychology*. Worcester, MA: Clark University Press, pp. 798–884.

Argyle, M. and Cook, M. 1976. *Gaze and Mutual Gaze*. Cambridge: Cambridge University Press.

Atkinson, M. 1984. *Our Masters' Voices: The Language and Body Language of Politics*. London: Methuen.

Auburn, T., Drake, S., and Willig, C. 1995. You punched him, didn't you? Versions of violence in accusatory interviews. *Discourse & Society*, 6, 353–386.

Austin, J.L. 1961. *Philosophical Papers*. Oxford: Clarendon.

Barber, N., Taylor, C., and Strick, S. 2009. Wine consumers' environmental knowledge and attitudes: Influence on willingness to purchase. *International Journal of Wine Research*, 1, 59–72.

Barthes, R. 1981. *Camera Lucida: Reflections on Photography*. New York: Hill & Wang.

Bartlett, F.C. 1932. *Remembering: A Study in Experimental and Social Psychology*. Cambridge: Cambridge University Press.

Baumeister, R.F., Vohs, K.D., and Funder, D.C. 2007. Psychology as the science of self-reports and finger movements: Whatever happened to actual behavior? *Perspectives on Psychological Science*, 2, 396–403.

Beattie, G. 1992. *We are the People: Journeys through the Heart of Protestant Ulster*. London: Heinemann.

Beattie, G. 2004. *Visible Thought: The New Psychology of Body Language*. London: Routledge.

Beattie, G. 2010. *Why Aren't We Saving the Planet? A Psychologist's Perspective*. London: Routledge.

Beattie, G. 2013. *Our Racist Heart? An Exploration of Unconscious Prejudice in Everyday Life*. London: Routledge.

Beattie, G. 2016. *Rethinking Body Language. How Hand Movements Reveal Hidden Thoughts*. London: Routledge.

Beattie, G. 2018a. *The Conflicted Mind: And Why Psychology Has Failed to Deal with It*. London: Routledge.

Beattie, G. 2018b. *The Body's Little Secrets*. London: Gibson Square.

Beattie, G. 2018c. Optimism bias and climate change. *The British Academy Review*, 33, 12–15.

Beattie, G. and Ellis, A. 2017. *The Psychology of Language and Communication: Psychology Press Classic Editions.* London: Routledge.
Beattie, G. and McGuire, L. 2015. Harnessing the unconscious mind of the consumer: How implicit attitudes predict pre-conscious visual attention to carbon footprint information on products. *Semiotica*, 204, 253–290.
Beattie, G. and McGuire, L. 2016. Consumption and climate change. Why we say one thing but do another in the face of our greatest threat. *Semiotica*, 213, 493–538.
Beattie, G. and McGuire, L. 2018. *The Psychology of Climate Change.* London: Routledge.
Beattie, G. and McGuire, L. 2020. Reading the signs of climate change: Does dispositional optimism affect climate change mitigation? *Semiotica.*
Beattie, G., Cohen, D., and McGuire, L. 2013. An exploration of possible unconscious ethnic biases in higher education: The role of implicit attitudes on selection for university posts. *Semiotica*, 197, 171–201.
Beattie, G., Marselle, M., McGuire, L., and Litchfield, D. 2017. Staying over-optimistic about the future: Uncovering attentional biases to climate change messages. *Semiotica*, 218, 22–64.
Bechara, A. 2004. The role of emotion in decision-making: Evidence from neurological patients with orbitofrontal damage. *Brain and Cognition*, 55, 30–40.
Bechara, A., Damasio, H., Tranel, D., and Damasio, A.R. 1997. Deciding advantageously before knowing the advantageous strategy. *Science*, 275, 1293–1295.
Bechara, A., Tranel, D., and Damasio, H. 2000. Characterization of the decision-making impairment of patients with bilateral lesions of the ventromedial prefrontal cortex. *Brain*, 123, 2189–2202.
Beloff, H. 1985. *Camera Culture.* Oxford: Blackwell.
Black, I.R. and Morton, P. 2015. Appealing to men and women using sexual appeals in advertising: In the battle of the sexes, is a truce possible? *Journal of Marketing Communications*, 4, 331–350.
Braendel, K. 2009. Second chance luck. *Sporting Classics*, 28, 76–81.
Campbell, J., Schermer, J.A., Villani, V.C., Nguyen, B., Vickers, L., and Vernon, P.A. 2009. A behavioral genetic study of the Dark Triad of personality and moral development. *Twin Research and Human Genetics*, 12, 132–136.
Child, K.R. and Darimont, C.T. 2015. Hunting for trophies: Online hunting photographs reveal achievement satisfaction with large and dangerous prey. *Human Dimensions of Wildlife*, 20, 531–541.
Christie, R. and Geis, F.L. (Eds.). 1970. *Studies in Machiavellianism.* San Diego, CA: Academic Press.
Codding, B.F., Bliege-Bird, R., and Bird, D.W. 2011. Provisioning offspring and others: Risk–energy trade-offs and gender differences in hunter–gatherer foraging strategies. *Proceedings of the Royal Society of London: Biological Sciences*, 278, 2502–2509.
Corral-Verdugo, V. 1997. Dual 'realities' of conservation behavior: Self-reports versus observations of re-use and recycling behavior. *Journal of Environmental Psychology*, 17, 135–145.
Crysel, L.C., Crosier, B.S., and Webster, G.D. 2013. The Dark Triad and risk behavior. *Personality and Individual Differences*, 54, 35–40.

Damasio, A.R. 1994. *Descartes' Error: Emotion, Reason and the Human Brain.* New York: Putnam.

Darimont, C.T., Codding, B.F., and Hawkes, K. 2017. Why men trophy hunt. *Biology Letters* 13.20160909. http://rsbl.royalsocietypublishing.org/content/13/3/20160909

Dawkins, R. 1976. *The Selfish Gene.* Oxford: Oxford University Press.

Di Minin, E., Leader-Williams, N. and Bradshaw, C.J.A. 2016. Banning trophy hunting will exacerbate biodiversity loss. *Trends in Ecology and Evolution*, 31, 99–102.

Duclos, J. 2017. Is hunting moral? A philosopher unpacks the question. In *The Conversation.* https://theconversation.com/is-hunting-moral-a-philosopher-unpacks-the-question-68645 (Accessed 18 July 2019).

Ebeling-Schuld, A.M. and Darimont C.T. 2017. Online hunting forums identify achievement as prominent among multiple satisfactions. *Wildlife Society Bulletin*, 41, 523–529.

Edwards, D. and Potter, J. 1992. *Discursive Psychology.* London: Sage.

Ekman, P. 1985. *Telling Lies.* New York: Norton.

Ekman, P. 1989. The argument and evidence about universals in facial expressions. In H. Wagner and A. Manstead (Eds.). *Handbook of Social Psychophysiology.* Chichester: Wiley, pp. 143–164.

Ekman, P. 1992. An argument for basic emotions. *Cognition and Emotion*, 6, 169–200.

Ekman, P. and Friesen, W.V. 1969. The repertoire of nonverbal behavior: Categories, origins, usage, and coding. *Semiotica*, 1, 49–98.

Ekman, P. and Friesen, W.V. 1982. Felt, false, and miserable smiles. *Journal of Nonverbal Behavior*, 6, 238–252.

Ekman, P., Davidson, R.J., and Friesen, W.V. 1990. The Duchenne smile: Emotional expression and brain physiology: II. *Journal of Personality and Social Psychology*, 58, 342–353.

Elbert, T., Weierstall, R., and Schauer, M. 2010. Fascination violence: On mind and brain of man hunters. *European Archives of Psychiatry and Clinical Neuroscience*, 260, 100–105.

Exline, R.V. and Winters, L.C. 1965. Affective relations and mutual glances in dyads. In S.S. Tomkins and C.E. Izard (Eds.), *Affect, Cognition, and Personality.* New York: Springer, pp. 319–350.

Fehr, B., Samson, B., and Paulhus, D.L. 1992. The construct of Machiavellianism: Twenty years later. In C.D. Spielberger and J.N. Butcher (Eds.), *Advances in Personality Assessment.* Hillsdale, NJ: Erlbaum, pp. 77–116.

Fielding, K.S., van Kasteren, Y., Louis, W., McKenna, B., Russell, S., and Spinks, A. 2016. Using individual householder survey responses to predict household environmental outcomes: The cases of recycling and water conservation. *Resources, Conservation and Recycling*, 106, 90–97.

Frank, M.G. and Ekman, P. 1993. Not all smiles are created equal: The differentiation between enjoyment and non-enjoyment smiles. *Humor*, 6, 9–26.

French, V. 2008. *The Bag of Bones.* London: Walker Books.

Furnham, A., Richards, S.C., and Paulhus, D.L. 2013. The Dark Triad of personality: A 10-year review. *Social and Personality Psychology Compass* 7, 199–216.

Garry, M. and Gerrie, M.P. 2005. When photographs create false memories. *Current Directions in Psychological Science*, 14, 321–325.

Graham, J.A. and Argyle, M. 1975. A cross-cultural study of the communication of extra-verbal meaning by gestures. *International Journal of Psychology*, 10, 57–67.

Greenwald, A.G., Nosek, B.A., and Banaji, M.R. 2003. Understanding and using the Implicit Association Test: I. An improved scoring algorithm. *Journal of Personality and Social Psychology*, 85, 197–216.

Griskevicius, V., Tybur, J.M., Sundie, J.M., Cialdini, R.B., Miller, G.F., and Kenrick, D.T. 2007. Blatant benevolence and conspicuous consumption: When romantic motives elicit strategic costly signals. *Journal of Personality and Social Psychology*, 93, 85–102.

Griskevicius, V., Tybur, J.M., and Van den Bergh, B. 2010. Going green to be seen: Status, reputation, and conspicuous conservation. *Journal of Personality and Social Psychology*, 98, 392–404.

Gullone, E. and Robertson, N. 2008. The relationship between bullying and animal abuse behaviors in adolescents: The importance of witnessing animal abuse. *Journal of Applied Developmental Psychology*, 29, 371–379.

Gurven, M., Allen-Arave, W., Hill, K., and Hurtado, M. 2000. It's a wonderful life: Signaling generosity among the Ache of Paraguay. *Evolution and Human Behavior*, 21, 263–282.

Gustafson, S.B. and Ritzer, D.R. 1995. The dark side of normal: A psychopathy-linked pattern called aberrant self-promotion. *European Journal of Personality*, 9, 147–183.

Haidt, J. 2001. The emotional dog and its rational tail: A social intuitionist approach to moral judgment. *Psychological Review*, 108, 814.

Halkowski, T. 1990. Role as an interactional device. *Social Problems*, 37, 564–577.

Hare, R.D. 1985. Comparison of procedures for the assessment of psychopathy. *Journal of Consulting and Clinical Psychology*, 53, 7–16.

Hart, S. and Hare, R.D. 1998. Association between psychopathy and narcissism: Theoretical views and empirical evidence. In E.F. Ronningstam (Ed.), *Disorders of Narcissism: Diagnostic, Clinical, and Empirical Implications*. Washington: American Psychiatric Press, pp. 415–436.

Hawkes, K. 1991. Showing off: Tests of a hypothesis about men's foraging goals. *Ethology and Sociobiology*, 12, 29–54.

Hawkes, K. 1993. Why Hunter–Gatherers work. *Current Anthropology*, 34, 341–361.

Hawkes, K. and Bliege Bird, R. 2002. Showing off, handicap signaling, and the evolution of men's work. *Evolutionary Anthropology: Issues, News, and Reviews*, 11, 58–67.

Heffetz, O. 2011. A test of conspicuous consumption: Visibility and income elasticities. *Review of Economics and Statistics*, 93, 1101–1117.

Honkanen, P., Verplanken, B., and Olsen, S.O. 2006. Ethical values and motives driving organic food choice. *Journal of Consumer Behaviour: An International Research Review*, 5, 420–430.

Isaacowitz, D.M. 2006. Motivated gaze: The view from the gazer. *Current Directions in Psychological Science*, 15, 68–72.

Jefferson, G. 1985. An exercise in the transcription and analysis of laughter. In T. Van Dijk (Ed.), *Handbook of Discourse Analysis*. London: Academic Press, pp. 25–34, vol 3.

Johnson, P.J., Kansky, R., Loveridge, A.J., and Macdonald, D.W. 2010. Size, rarity and charisma: Valuing African wildlife trophies. *PLOS ONE*, 5(9), e12866.

Johnston, L., Miles, L., and Macrae, C.N. 2010. Why are you smiling at me? Social functions of enjoyment and non-enjoyment smiles. *British Journal of Social Psychology*, 49, 107–127.

Jonason, P.K., Lyons, M., Bethell, E.J., and Ross, R. 2013. Different routes to limited empathy in the sexes: Examining the links between the Dark Triad and empathy. *Personality and Individual Differences*, 54, 572–576.

Jonason, P.K. and Schmitt, D.P. 2012. What have you done for me lately? Friendship-selection in the shadow of the Dark Triad traits. *Evolutionary Psychology*, 10, 192–199.

Jonason, P.K., Slomski, S., and Partyka, J. 2012. The Dark Triad at work: How toxic employees get their way. *Personality and Individual Differences*, 52, 449–453.

Jonason, P.K. and Webster, G.D. 2012. A protean approach to social influence: Dark Triad personalities and social influence tactics. *Personality and Individual Differences*, 52, 521–526.

Jordan, R. and Beattie, G. 2003. Understanding male interpersonal violence: A discourse analytic approach to accounts of violence on the door. *Semiotica*, 144, 101–142.

Kahneman, D. 2011. *Thinking Fast and Slow*. New York: Farrar, Straus and Giroux.

Kaplan, H. 1983. The evolution of food sharing among adult conspecifics: Research with Ache hunter–gatherers of Eastern Paraguay. *Doctoral Dissertation*. University of Utah, Salt Lake City. http://permalink.opc.uva.nl/item/000489616

Kavanagh, P.S., Signal, T.D., and Taylor, N. 2013. The Dark Triad and animal cruelty: Dark personalities, dark attitudes, and dark behaviors. *Personality and Individual Differences*, 55, 666–670.

Kelley, H. 1967. Attribution theory in social psychology. *Nebraska Symposium on Motivation*, 15, 192–238.

Kelly, J.R. and Rule, S. 2013. The hunt as love and kill: Hunter–prey relations in the discourse of contemporary hunting magazines. *Nature and Culture*, 8, 185–204.

Kerasote, T. 2007. *Merle's Door: Lessons from a Freethinking Dog*. New York: Harcourt.

Khan, R. and Cooke, D.J. 2008. Risk factors for severe inter-sibling violence: A preliminary study of a youth forensic sample. *Journal of Interpersonal Violence*, 23, 1513–1530.

Koenigs, M., Young, L., Adolphs, R., Tranel, D., Cushman, F., Hauser, M., and Damasio, A. 2007. Damage to the prefrontal cortex increases utilitarian moral judgments. *Nature*, 446, 908–911.

Kormos, C. and Gifford, R. 2014. The validity of self-report measures of pro-environmental behavior: A meta-analytic review. *Journal of Environmental Psychology*, 40, 359–371.

LaFrance, M. and Hecht, M.A. 1995. Why smiles generate leniency. *Personality and Social Psychology Bulletin*, 21, 207–214.

Leathwood, C., Maylor, U., and Moreau, M. 2009. *The Experience of Black and Minority Ethnic Staff Working in Higher Education*. London: Equalities Challenge Unit.

Lee, V. and Beattie, G. 1998. The rhetorical organization of verbal and nonverbal behavior in emotion talk. *Semiotica*, 120, 39–92.

Lee, V. and Beattie, G. 2000. Why talking about negative emotional experiences is good for your health: A micro-analytic perspective. *Semiotica*, 130, 1–82.

Lee, Y. 1993. Recycling behavior and waste management planning. *Journal of Building and Planning*, 7, 65–77.

Leopold, A. 1933. *Game Management*. New York: Scribner's.

Leviston, Z. and Walker, I. 2012. Beliefs and denials about climate change: An Australian perspective. *Ecopsychology*, 4, 277–285.

MacDonald, J.D. 1987. *A Deadly Shade of Gold*. Boston: G.K. Hall.

Macdonald, J.M. 1963. The threat to kill. *American Journal of Psychiatry*, 120, 125–130.

Mallett, X. 2015. Why we may never understand the reasons people hunt animals as 'trophies'. *The Conversation*. https://theconversation.com/why-we-may-never-understand-the-reasons-people-hunt-animals-as-trophies-45701 (Accessed 6 July 2018).

Markham B. 1942. *West with the Night*. Boston: Houghton.

Marshall, G. 2015. *Don't Even Think About It: Why Our Brains Are Wired to Ignore Climate Change*. London: Bloomsbury Publishing.

Maynard, M. 2007. Say 'Hybrid' and many people will hear 'Prius'. *The New York Times*. https://www.nytimes.com/2007/07/04/business/04hybrid.html (Accessed 24 August 2018).

McGrath, B. 2014. An interview with Kendall Jones. *First For Hunters*. https://firstforhunters.wordpress.com/2014/07/21/an-interview-with-kendall-jones/ (Accessed 21 July 2018).

McHoskey, J. 1995. Narcissism and Machiavellianism. *Psychological Reports*, 77, 755–759.

Meere, M. and Egan, V. 2017. Everyday sadism, the Dark Triad, personality and disgust sensitivity. *Personality and Individual Differences*, 112, 157–161.

Mehrabian, A. and Friar, J.T. 1969. Encoding of attitude by a seated communicator via posture and position cues. *Journal of Consulting and Clinical Psychology*, 33, 330–336.

Nelson, M.P., Bruskotter, J.T., Vucetich, J.A., and Chapron, G. 2016. Emotions and the ethics of consequence in conservation decisions: Lessons from Cecil the Lion. *Conservation Letters*, 9, 302–306.

Olson, M. 1981. Consumer attitudes toward energy conservation. *Journal of Social Issues*, 37, 108–131.

Paulhus, D.L. and Williams, K.M. 2002. The Dark Triad of personality: Narcissism, Machiavellianism, and psychopathy. *Journal of Research in Personality*, 36, 556–563.

Pendergrass, W.S., Payne, C.A., and Buretz, G.R. 2016. Cybershaming: The shallowing hypothesis in action. *Issues in Information Systems*, 17, 65–75.

Pennebaker, J.W. 2000. Telling stories: The health benefits of narrative. *Literature and Medicine*, 19, 3–18.

Peterson, N. 2004. An approach for demonstrating the social legitimacy of hunting. *Wildlife Society Bulletin*, 32, 310–321.

Potter, J. 1996. *Representing Reality: Discourse, Rhetoric and Social Construction*. London: Sage.

Potter, J. and Wetherell, M. 1987. *Discourse and Social Psychology: Beyond Attitudes and Behaviour*. London: Routledge.

Raskin, R.N. and Terry, H. 1988. A principal-components analysis of the Narcissistic Personality Inventory and further evidence of its construct validity. *Journal of Personality and Social Psychology*, 54, 890–902.

Rathje, W. 1989. The three faces of garbage. Measurements, perceptions, behaviors. *Journal of Management and Technology*, 17, 61–65.

Ross, L. 1977. The intuitive psychologist and his shortcomings: Distortions in the attribution process. In L. Berkowitz (Ed.), *Advances in Experimental Social Psychology*. New York: Academic Press, pp. 173–220.

Rust, N. and Verissimo, D. 2015. Why killing lions like Cecil may actually be good for conservation. In *The Conversation*. https://theconversation.com/why-killing-lions-like-cecil-may-actually-be-good-for-conservation-45400 (Accessed 18 July 2019).

Rutter, D.R. and Stephenson, G.M. 1979. The role of visual communication in social interaction. *Current Anthropology*, 20, 124–125.

Sacks, O. 1984. On doing 'being ordinary'. In J.M. Atkinson and J. Heritage (Eds.), *Structures of Social Action: Studies in Conversation Analysis*. Cambridge: Cambridge University Press, pp. 413–429.

Schlegelmilch, B.B., Bohlen, G.M., and Diamantopoulos, A. 1996. The link between green purchasing decisions and measures of environmental consciousness. *European Journal of Marketing*, 30, 35–55.

Scott, M.B. and Lyman, S.M. 1968. Accounts. *American Sociological Review*, 33, 46–62.

Shepherd, J. 2011. 14,000 British Professors – But Only 50 Are Black. *The Guardian*. https://www.theguardian.com/education/2011/may/27/only-50-black-british-professors (Accessed 18 July 2019).

Simons, D.A., Wurtele, S.K., and Durham, R.L. 2008. Developmental experiences of child sexual abusers and rapists. *Child Abuse & Neglect*, 32, 549–560.

Steinbeck, J. 1951. *The Log from the Sea of Cortez*. New York: Penguin.

Storey, W.K. 2008. *Guns, Race, and Power in Colonial South Africa*. New York: Cambridge University Press.

Tsakiridou, E., Boutsouki, C., Zotos, Y., and Mattas, K. 2008. Attitudes and Behaviour towards organic products: An exploratory study. *International Journal of Retail & Distribution Management*, 36, 158–175.

Tversky, A. and Kahneman, D. 1973. Availability: A heuristic for judging frequency and probability. *Cognitive Psychology*, 5, 207–232.

Twenge, J.M. and Campbell, K.W. 2009. *The Narcissism Epidemic. Living in the Age of Entitlement*. New York: Atria.

US Department of Education. 2007. https://nces.ed.gov/pubs2007/2007064.pdf (Accessed 27 February 2019).

Veblen, T. 1899. *The Theory of the Leisure Class*. New York: New American Library.

Volant, A.M., Johnson, J.A., Gullone, E., and Coleman, G.J. 2008. The relationship between domestic violence and animal abuse: An Australian study. *Journal of Interpersonal Violence*, 23, 1277–1295.

Walsh, D. and Gentile, D. 2007. Slipping under the radar: Advertising and the mind. In L. Riley and I. Obot (Eds.) *Driving It In: Alcohol Marketing and Young People*. Geneva: World Health Organization.

Warfield, J. 2019. Creepiest Festival for Trophy Hunters Is Kicking off This Week. *The Dodo*. https://www.thedodo.com/in-the-wild/trophy-hunting-convention-nevada (Accessed 7 May 2019).

Warriner, G.K., McDougall, G.H., and Claxton, J.D. 1984. Any data or none at all? Living with inaccuracies in self-reports of residential energy consumption. *Environment and Behavior*, 16, 503–526.

Wiessner, P. 2002. Hunting, healing, and hxaro exchange: A long-term perspective on Kung (Ju/'hoansi) large-game hunting. *Evolution and Human Behavior*, 23, 407–436.

Winterhalter, B. 1986. Diet choice, risk and food sharing in a stochastic environment. *Journal of Anthropological Archaeology*, 5, 369–392.

Zahavi, A. and Zahavi, A. 1997. *The Handicap Principle: A Missing Piece of Darwin's Puzzle*. New York: Oxford University Press.

Index